Advances in Reactor Measurement and Control

Advances in Reactor Measurement and Control

by Gregory K. McMillan

ISA™

Notice

The information presented in this publication is for the general education of the reader. Because neither the author nor the publisher has any control over the use of the information by the reader, both the author and the publisher disclaim any and all liability of any kind arising out of such use. The reader is expected to exercise sound professional judgment in using any of the information presented in a particular application.

Additionally, neither the author nor the publisher has investigated or considered the effect of any patents on the ability of the reader to use any of the information in a particular application. The reader is responsible for reviewing any possible patents that may affect any particular use of the information presented.

Any references to commercial products in the work are cited as examples only. Neither the author nor the publisher endorses any referenced commercial product. Any trademarks or tradenames referenced belong to the respective owner of the mark or name. Neither the author nor the publisher makes any representation regarding the availability of any referenced commercial product at any time. The manufacturer's instructions on use of any commercial product must be followed at all times, even if in conflict with the information in this publication.

Printed in the United States of America.
10 9 8 7 6 5 4 3 2

ISBN 978-0-87664-089-0

ISA
67 Alexander Drive
P.O. Box 12277
Research Triangle Park, NC 27709

Library of Congress Cataloging-in-Publication Data in process.

Acknowledgments

Publications by Bill Luyben and Greg Shinskey are the source of most of my deep knowledge in reactor control, particularly in dealing with recycles effects and open loop unstable processes. Most of the rest of my knowledge has been gained by working with and simulating continuous fluidized bed reactors for monomers and batch reactors for polymers, specialty chemicals, ethanol, and biologics.

There were many Monsanto and Solutia Directors and Senior Fellows who played a key role, technically and personally in my journey as an automation professional. Foremost among those who shaped my career was the late Dr. Jim Fair, Honorary Professor Emeritus at the University of Texas, the creator and first director of the Engineering Technology (ET) Department. Being part of the ET Department allowed access to the best technical minds in the world, along with the freedom to explore and the encouragement to publish. My second ET director, Dr. Stan Proctor, shared his extensive knowledge in modeling and control and provided direction and support. The appreciation of my independence and humor by my last ET director, Dr. Adrian Hutton, another Texan, enabled me to publish many funny and creative articles and books. The knowledge I inherited from the late Vernon Trevathan and gained from conversations with Senior Fellows at Solutia and Monsanto, including Dr. Henry Chien, Dr. Bob Otto, Dr. Terry Tolliver, and Stan Weiner, has been of inestimable value.

When I retired from Solutia, I was fortunate to have the opportunity to become a consultant to Emerson Process Management at their headquarters in Austin in 2003. Here I have had access to some of the best minds on the supplier side, including James Beall, Terry Blevins, Mark Coughran, Lou Heavner, Mike Ott, and Willy Wojsznis. Jim Cahill has helped me to effectively use social media. The support by Mark Nixon, manager of DeltaV R&D, and Grant Wilson, vice president of DeltaV technology of my endeavors has enabled me to continue my journey. Most recently my role as a simulation consultant at MYNAH Technologies, LLC has opened the door to seeing to fruition the development of a bioreactor model that can be deployed to gain better process understanding and recognition of control opportunities. Martin

Berutti, Nobin William, and Dave Sextro at MYNAH have created an inspiring and friendly environment to promote creativity and achievement, capitalizing on the talent and enthusiasm of the next generation of engineers to get the most out of simulation.

Special thanks go to Pierce Wu at MYNAH Technologies, LLC who developed in MiMiC the linear dynamic estimators for self-regulating, integrating, and runaway processes in Chapter 2 that can be used as inferential measurements. Pierce demonstrated that these estimators could be readily adapted online to match the dynamic response of the process.

I personally thank Héctor Torres, a coauthor of papers and principal participant in the ISA Mentor program, for his effective and timely review of the material in this book on an extremely aggressive schedule. This book came together in just one month. Héctor's inputs have resulted in many improvements and clarifications.

The extraordinary effort by Susan Colwell, manager of Publications Development, ISA, on evenings and weekends has enabled the ISA Mentor Program to be a success and this book to become a reality. Finally, the patience and understanding of my wife, Carol, have made this book and the sharing of what I have learned a part of our life.

Contents

Preface

The reactor sets the stage for plant capability. Reactor performance largely determines product quality and process capacity. The product coming out of the reactor has to meet product specifications or production requirements to give downstream processes for separation and purification a chance to do their job. Advances in the use of control strategies, inferential measurements, and PID tuning and features for the different types of reactors have increased the opportunities for control and optimization.

There is an incredible difference in process dynamics and tuning rules based on the type of reactor. Considerable disagreement has consequently arisen as to the tuning settings and the importance of the dynamics of automation system components. Each control strategy is considered special and improvements by simple configuration changes are not recognized.

The implications of the usage of different types of reactors and the subsequent process dynamics are illustrated when considering controlling temperature by manipulating reactant feed. While inherently maximizing production rate, this strategy is assiduously avoided in liquid reactors but is effectively utilized in fluidized bed gas reactors; there is essentially no inverse response or process time constant in these gas reactors. The process dead time, the residence time (volume/flow), is a few seconds for the high flow rate in these high capacity processes. The largest time constant in the loop is the temperature sensor, which leads to possible attenuation and deception. Other gas flow reactors have recycle streams and heat integration that introduce complex and fast dynamics. The tuning objective for these reactors is to minimize sudden changes and overshoot in the PID output rather than in disturbance rejection. PID gains in the range of 1 to 4 are used, as larger sensor lags offer the ability to increase the PID gain and reduce the amplitude of process oscillations, giving an illusion of better control. This illusion can lead to product quality problems and an increase in the lag by sensor selection, installation, and maintenance practices.

Liquid well-mixed continuous reactors have completely different dynamics and tuning. The process time constant is the residence time. The process time constant is large, approaching an hour for slow reactions. The process dead time is ½ of the turnover time and is generally just a few seconds. In these reactors, the process time constant is 100 times or more the process dead time. The tuning rules minimize excursions of the process variable, maximizing disturbance rejection. The PID output is changed aggressively to minimize errors in the PID input. Overshoot of the PID output is necessary to reach setpoint in a reasonable time frame with PID gains of 10 to 40 being common. The process response is approximated as near-integrating, and the integrating process tuning rules developed for batch processes are used.

Batch processes have an extremely slow response and even more aggressive changes in the PID output are employed. Overshoot is necessary for the PID input to reach a new setpoint. PID gains of 20 to 80 are used for tight control. Runaway processes require high gains for stability.

A key feature to enable tight control without upsetting utility systems from aggressive changes in the PID output is external reset feedback, while rate limits on the setpoint being manipulated by the PID output provide directional move suppression. This PID feature will be shown to be helpful in many other ways for control and optimization of all types of reactors.

Despite the importance of reactor control and the wide variety of process responses, which leads to a considerable spectrum of control strategies and tuning objectives, this is the first book dedicated to the subject of reactor measurement and control from a practitioner's perspective. Concepts are developed to help the reader understand the fundamental differences in reactor applications and provide the ability to improve the performance of nearly all types of reactors.

1
System Dynamics

1
System Dynamics

The types of process responses encountered in industry are defined based on an open loop test where the PID controller is in manual or remote output, so there is no feedback correction by the PID to the process (i.e., there is no closed loop response). This test removes the effect of the PID algorithm and tuning in defining the process response. Tests are done for step changes in the controller output both above and below the normal output. The process response is observed until the process can be identified. For processes that slowly change or have no steady state, the identification of the dead time and the subsequent maximum ramp rate in the right direction is sufficient. For processes that can runaway (e.g., highly exothermic reactors), an open loop test must be terminated before the process starts to accelerate.

For processes where the ramp rate or acceleration is potentially too fast, the controller must never be in manual mode. For these and other systems where the feedback control must always be active, the identification of the open loop dynamics is done with the controller in automatic mode by step changes in the setpoint or by pulses injected into the PID output. These pulses are step changes that are held long enough to identify the response. During the test, there should be no disturbances to the system, so that the process response seen is entirely the result of the step changes in PID output or setpoint.

> ***An open loop response (response without feedback action) for a step change in PID output is used to define a control loop's dynamics. For loops that must stay in automatic mode, the open loop response is computed from a step change in the PID setpoint or a pulse injected into the PID output.***

The response observed in these tests includes: response of the analog output, final control element (e.g., control valve or variable frequency drive), process, sensor, transmitter, analog input, and process variable (PV) input to the PID. The observed response includes the effect of velocity limits, dead times, time constants, and gains in the automation system. Better terminology would be "open loop response" than "process response" because the observed response includes almost everything in the loop response. Also, a qualifier that designates the source of the dynamics should precede the particular dynamic term (e.g., valve dead time, process dead time, and measurement dead time).

> ***The open loop response includes the dynamics of the automation system and process. Qualifiers should be added to the terms that contribute to the open loop response to identify the source of the dynamics.***

1-1. Sources of Loop Dynamics

Figure 1.1 shows the types and locations of dynamics in a reactor control loop. In this figure and elsewhere in the book, "lag" is used as the shorter term for "time constant" and "delay" as the shorter term for "dead time". Figures, text, and tables will use these terms interchangeably.

All processes have multiple time constants in the response of process and automation system components. A time constant is the time required for the change in component output to reach 63% of its final value after the output starts to respond to a step change in the input to the component. In reactors, the process time constants are primarily associated with heat transfer and well-mixed volumes. Temperature measurement time constants originate from sensor, thermowell and transmitter lags. The controller time constant is the result of filters on the analog input or PID process variable.

Processes with a steady state are termed self-regulating processes because these processes have internal negative feedback that will result in the process output variable settling out at a final value for a change in a process input. A disturbance or a change in a PID output that becomes a change in a manipulated variable, which is in most cases is a change in flow, is the source of the change in the process input. For self-regulating processes (e.g., flow and continuous composition, pH, and temperature control of small volumes), a first order plus dead time or second order plus dead time approximation is used. In a first order plus time dead time model (FOPDT), the largest time constant is identified along with the open loop steady state gain and total dead time. In

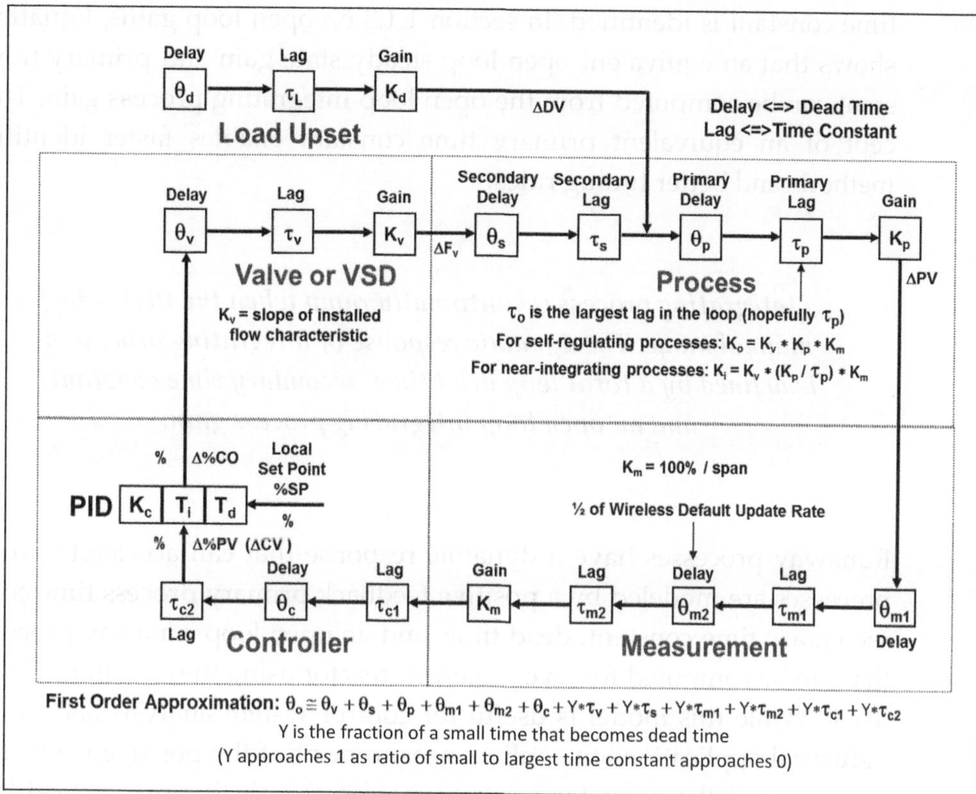

Figure 1.1 – Types and Locations of Dynamics in the PID Control Loop

a second order plus dead time model (SOPDT), the second largest time constant is also identified. The largest time constant is called the primary time constant and the second largest time constant is called the secondary time constant. In this book, the largest time constant is also termed the open loop time constant, signifying that the response is for the PID in manual mode (the loop is open).

> ***Self-regulating processes have an observable steady state when the PID controller is in manual mode. The dynamic response of self-regulating processes is defined by a total loop dead time, primary and secondary time constants, and an open loop steady state gain.***

For processes where a steady state does not exist or is not attainable in a reasonable time frame, an open loop integrating process gain, dead time, and a time constant is identified. In section 1.1.3 on open loop gains, Equation 1.3a shows that an equivalent open loop steady state gain and primary time constant can be computed from the open loop integrating process gain. The concept of an equivalent primary time constant enables faster identification methods and better tuning rules.

Integrating processes continually ramp when the PID is in manual mode. The dynamic response of integrating processes is defined by a total loop dead time, secondary time constant, and an open loop integrating process gain.

Runaway processes have a dynamic response that can accelerate. Runaway processes are modeled by a positive feedback primary process time constant, secondary time constant, dead time, and an open loop runaway process gain that can be computed for an exothermic reactor using the equations in Appendix F. While this model is useful for control system analysis and design, in industrial applications the online measurement of the positive feedback time constant and the open loop runaway process gain is not practical because acceleration is prevented for safety reasons.

Runaway processes will continually move and eventually accelerate when the PID is in manual mode. The dynamic response of runaway processes is defined by a total loop dead time, positive feedback primary time constant, secondary time constant, and an open loop runaway process gain. In industrial systems, runaway processes are treated as integrating processes.

1.1.1 Dead Times

All processes have multiple sources of dead times that result in a total dead time, which is defined as the time interval between the step change in PID output and the first recognizable change in the PID input. Noise can delay the signal recognition until the excursion is beyond the noise band, thus creating a longer dead time. The observed dead time is frequently called the process dead time. The observed dead time is really a total loop dead time (θ_o) that is the sum of all the pure dead times and the equivalent dead times from all time constants smaller than the largest time constant in the loop for a first order (one time constant) plus dead time approximation. For a second order plus dead time approximation that includes a secondary time constant, all time constants smaller the secondary time constant create an equivalent dead time. The secondary time constant creates the bend in the initial response right after the dead time. The equivalent dead time increases from 30% to 99% of the time constant as the ratio of the time constant to the largest time constant gets smaller. Time constants that are small compared to the largest time constant are summed as being essentially 100% dead time to help make up for inevitable unknown time constants.

Appendix C develops equations for loop performance that show the ultimate limit to the peak and integrated errors for load disturbances are proportional to the dead time and dead time squared, respectively. The equations for tuning in this chapter show that the maximum PID gain for stability is infinite if the dead time is zero (assuming no noise). At the end of this chapter Equations 1.20a, 1.20b and 1.20c show that the minimum time to reach a new setpoint is the dead time and no PID output or tuning limits.

> ***If there was no dead time and no PID limits, perfect control would be possible. Errors from disturbances would be zero and a new setpoint could be reached instantaneously.***

For processes with negligible back mixing, such as fluidized bed gas reactors, plug flow reactors, static mixers, pipes, tubes, coils and jackets, most of the residence time (volume/flow) becomes a detrimental transportation delay. For processes with axial mixing due to turbulence, recirculation, or agitation, most of the residence time becomes a beneficial process time constant.

> ***An increase in residence time at low production rates will increase the process dead time from volumes with negligible back mixing.***

The dead time from analyzers is 1.5 times the analyzer cycle time plus the sample transportation delay. The dead time from controller execution rates, digital scan rates or update rates including wireless default update rates is ½ the update time or rate plus any latency. For most digital devices calculations, the latency is zero. The exception is some of the older model predictive controllers that were computationally intensive. Note that the execution, scan, or update rate is a time interval (e.g., seconds per update) rather than a frequency (e.g., updates per second).

Not seen in the open loop responses, defined by a step change in PID output, is the additional delay experienced in closed loop operation resulting from the time it takes for the process variable to exceed the measurement threshold sensitivity or resolution limit and the output signal to pass through the deadband (backlash), threshold sensitivity (stiction), and resolution limits of the final control element. The dead time from these non-ideal characteristics of the automation system is not seen when the test uses a step change larger than these limits. In closed loop operation, a step change in output can occur for a step change in setpoint but subsequent closed loop action to recover from overshoot or disturbances involves gradual changes in the output. The additional dead time from these limits can be approximated as the limit divided by the rate of change of the signal (e.g., valve deadband or resolution limit divided by the rate of change of the PID output).

1.1.2 Time Constants

The open loop time constant (τ_o) is the largest time constant in the open loop response for a step change in a manual PID output. In contrast, the closed loop time constant is the largest time constant in the closed loop response for a step change in an automatic PID setpoint. The closed loop time constant is the tuning parameter for Internal Model Control (IMC) and lambda tuning methods for self-regulating processes, which focus on setpoint response. For processes where disturbances that enter downstream of the process or where the primary process time constant is not significantly larger than the total dead time, the tuning for disturbances for setpoint changes is similar for the most common PID Form and Structure without a setpoint filter or lead-lag.

While often called the process time constant, the largest time constant can occur anywhere in the loop. For liquid pressure and flow control or concentration, pH, or temperature control of plug flow volumes without thermal lags, the largest time constant is usually somewhere in the automation system (e.g., valve, variable speed drive, sensor, transmitter, or DCS). Ideally, the largest time constant called the primary time constant of the process is downstream of where the disturbances and manipulated flow enter the process. We will see in Section 1.3 on tuning and Section 1.4 on loop performance, the effect of such a primary time constant is beneficial. Appendix C shows the ultimate limits as to how small the peak and integrated errors are inversely proportional to this primary time constant. The maximum allowable PID gain is proportional to the primary time constant. The primary process time constant slows down an excursion from a disturbance giving time for the PID to catch up with the upset by a correction in the PID output that stops the excursion.

> ***The existence of a large process time constant slows down the excursion from a load disturbance and allows a large PID gain that enables the PID to catch up and correct for the disturbance before the process variable gets too far from setpoint.***

If the largest time constant is in the measurement, the previously mentioned relationships are true but the measured process variable is not moving as fast as the actual process variable and the amplitude of an oscillation is attenuated by the filtering effect of the measurement time constant. The variability of the measurement on a trend chart will look much smaller than the variability of the actual process variable. The main clue as to whether a measurement time constant has become the largest time constant is an increase in the total loop dead time and a tendency to oscillate, with a larger period from the previous largest time constant now contributing to the loop dead time. For a runaway reaction, equations will be developed to show an unstable situation develops long before the measurement time constant approaches the primary process time constant.

The location and size of the largest time constant relative to the process dead time is the source of most of the disagreement in process control. The IMC and lambda tuning methods and most of the control literature are based on a primary process time constant that is not significantly larger than the total loop dead time. For a time constant that is between ¼ and 4 times the dead time (moderate self-regulating processes), tuning methods such as IMC and

> ***If the largest time constant is in the measurement, a serious deception develops where the trend chart looks better but the process is doing worse. The clue to this problem is an increase in loop dead time, slow oscillations and a decrease in product quality.***

lambda, which focus on a setpoint response, smooth gradual changes in the PID output and minimizing interactions, also give reasonable results for disturbances at the process input. Continuous processes without large well-mixed liquid volumes such as inline unit operations of pulp and paper and gas unit operations in the front end of refineries and hydrocarbon processes fit this criterion. Large continuous liquid processes in the chemical industry have time constant to dead time of 100 or more resulting in extremely poor load disturbance response for these methods. The solution for lambda tuning turned out to be a classification of the process response to be "near-integrating" with a simple switch to tuning rules for integrating processes when the time constant to dead time ratio exceeds 4, with a low limit of 4 dead times enforced on the reset time for PI control and a low limit of ½ the dead time enforced on the rate time for PID control. The "near-integrating" classification has been taught in courses in lambda tuning and used by specialists in lambda tuning for over a decade; however, lack of publications focusing on this essential practice has led to a pervasive misunderstanding as to suitability of lambda tuning for processes with a large time constant to dead time ratio.

The lambda for integrating process is an arrest time (time to stop a process excursion after a load disturbance). Self-regulating processes with a larger time constant to dead time ratio are termed near-integrating. The response of these near-integrating processes is a ramp similar to what is seen for a true integrating process in the time frame of the major reaction by the PID (e.g., 2 dead times). The response of runaway processes is also a ramp in the PID time frame because the PID must take action to prevent the start of acceleration. The maximum PID gain settings of near-integrating, true integrating, and runaway processes are orders of magnitude larger than what is shown in the literature.

In an open loop test, one cannot discern the location of the largest time constant. Focused tests are needed to identify the time constant of an automation system component. The time constant of a control valve or variable speed

> ***The location and size of the primary time constant relative to the total loop dead time is the source of most of the disagreement in tuning methods and objectives.***

drive is very dependent upon the size of the step change in the PID output. The step size must be larger than the deadband or resolution limit of the valve or variable speed drive (VSD). The time constant gets larger for small steps due to the sensitivity limits of the actuator and positioner. The time constant gets larger for large steps due to the valve slew rate or VSD setpoint rate limits.

The time constant of a measurement sensor is best measured after insertion in test solutions at the same operating conditions as at the point of installation. The time constant of temperature sensors and glass pH electrodes depends upon the velocity and direction of the change. Temperature sensors need to be in the thermowell when inserted into a temperature bath because the thermowell design, in particular the air gap between the sensor and thermowell tip, greatly affects the time constant. The remaining time constants in the transmission and processing of the measurement are readily determined from the transmitter damping settings and the signal filter settings in the analog input block and PID block.

> ***The identification of a temperature or pH sensor time constant requires immersion in test solutions at the same operating conditions as at the point of installation in the plant.***

During a test for a sensor time constant, an 86% response time (time to 86% of the final change) should be measured, as it is defined as two time constants, if the dead time is negligible. Waiting for the 98% response time (four time constants) is problematic and less useful. Noise and differences in threshold sensitivity can cause dramatically different results. Also, many sensors such as pH electrodes have a slow response after 86% of the final value is reached, causing an estimation of the time constant that is unreasonably long. Most of the PID controller correction occurs before the 86% response time is reached. For similar reasons an 86% response time is the basis of the ISA Standard for control valve response testing (ANSI/ISA-75.25.01-2000 (R2010) - *Test Procedure for Control from Step Inputs*). The statement of "response time" without specifying the percent of final response is of little value.

The 86% response time (time to reach 86% of final response) should be used to quantify the dynamic response of automation system components.

The second largest time constant (secondary time constant) is always detrimental, particularly to processes without a steady state in the time frame of PID action. The use of derivative action is particularly effective in counteracting the lag in these processes by introducing a secondary time constant into the PID correction. Thus, at a minimum, the rate time should be larger than the time constant of the sensor in a thermowell at operating conditions. For a heat transfer surface for temperature control of a large well-mixed liquid volume, the secondary time constant is likely to originate from the heat transfer surface lag unless the fit of the sensor within the thermowell is poor, the thermowell is coated, or the tip is in a stagnant area (e.g., the tip is behind a baffle or in a nozzle).

1.1.3 Open Loop Gains

Since the omission of the sources, units, and types of open loop gains in the literature has led to misconceptions and misunderstandings, this section goes into considerable detail in giving the basic equations that are used to define factors contributing to the open loop gains.

The open loop gains define what the PID sees as a % change in PV ($\Delta\%PV$) for a % change in controller output ($\Delta\%CO$). The open loop steady state gain is the product of the final control element, flow ratio, process, and measurement gains. Consider a loop with a control valve. For small changes, the final control element gain (change in flow divided by the change in % PID output) is the slope of the valve's installed flow characteristic curve at the current operating point (valve position). The flow ratio gain is the inverse of the feed flow for composition, pH, and temperature control. For flow, level, and pressure control, the flow ratio gain is simply one. For small changes, the process gain is the slope of a plot of the process variable versus the ratio of the manipulated flow to feed flow near the setpoint. The measurement gain (change in % PID input divided by the change in process variable) is 100% divided by the measurement span. Since the PID algorithm uses % signals, the open loop steady state gain must end up being the change in % PID input divided by the % change in PID output and hence dimensionless, despite the fact that the PID block and graphics show the PID process variable, and in some DCS the PID output, in engineering units.

Since the PID algorithm works with percent input and output signals, the open loop steady state gain must be dimensionless and the open loop integrating process gain has units of 1/sec.

The open loop steady state gain for a self-regulating process is the final change in the percent process variable input ($\Delta\%PV$) after the process has reached a new steady state for a step change in the percent controller output ($\Delta\%CO$).

$$K_o = \frac{\Delta\%PV}{\Delta\%CO} \tag{1.1a}$$

The open loop steady state gain (K_o) is the product of the gains for the valve or VSD (K_v), flow ratio (K_r), process (K_p), and measurement gains (K_m). The product of the flow ratio gain and process gain defined here is most often simply expressed as a process gain:

$$K_o = K_v * K_r * K_p * K_m \tag{1.1b}$$

For small changes in flow, the valve or VSD gain is the slope of the installed flow characteristic near the current operating point that is the change in flow in engineering units (ΔF_v) divided by the change in controller output ($\Delta\%CO$):

$$K_v = \frac{\Delta F_v}{\Delta\%CO} \tag{1.1c}$$

For composition, temperature, and pH loops the flow ratio gain (K_r) is inversely proportional to the main process feed flow (F_p):

$$K_r = \frac{\Delta R}{\Delta F_v} = \frac{\Delta F_v / F_p}{\Delta F_v} = \frac{1}{F_p} \tag{1.1d}$$

The process gain is the change in the process variable in engineering units (ΔPV) divided by the change in ratio (ΔR):

$$K_p = \frac{\Delta PV}{\Delta R} \tag{1.1e}$$

For flow and pressure loops the product of the flow ratio and process gain simply becomes the change in process variable per change in flow (both in engineering units):

$$K_r * K_p = \frac{\Delta PV}{\Delta F_v} \tag{1.1f}$$

It is not well recognized that loops for continuous composition, pH, and temperature control have an open loop steady state gain that is inversely proportional to feed rate.

The measurement gain (K_m) is 100% divided by the calibration span of the transmitter:

$$K_m = \frac{100\%}{PV_{max} - PV_{min}} \tag{1.1g}$$

Decreasing the measurement span will increase the open loop gain via an increase in the measurement gain. The PID gain setting must be decreased to prevent oscillations if the setting was already near the maximum allowable gain.

For the upper (primary) loop in a cascade control system, the PID output becomes the setpoint of a lower (secondary) loop rather than the setpoint of a valve or VSD. In Equation 1.1a, the setpoint scale gain for the manipulation of a setpoint is substituted for the gain for the manipulation of a valve or VSD. The nonlinearity of the installed flow characteristic that depends largely upon pressure drop for a valve and on static pressure for a VSD has been eliminated from the view of the upper loop. The gain for manipulation of a setpoint is computed as detailed in Equation 1.1i as the span of the setpoint scale of the lower loop divided by 100%. The setpoint scale normally matches the measurement scale of the lower loop. If the DCS output scale is in percent, this match happens automatically. If the PID output scale is in engineering units, the user is responsible for setting the upper PID output scale to match the lower PID scale. The most common lower (secondary) loop is flow, in which case there is still the flow ratio gain (K_r) for composition, pH, and temperature loops.

For cascade control where an upper composition, pH, or temperature loop is manipulating a lower flow loop, the valve or VSD gain is replaced with a setpoint gain in the open loop steady state gain giving:

$$K_o = K_{sp} * K_r * K_p * K_m \tag{1.1h}$$

The setpoint gain is the span of the lower loop setpoint scale divided by 100%:

$$K_{sp} = \frac{(SP_{max} - SP_{min})}{100\%} \tag{1.1i}$$

An important higher cascade control loop is an upper vessel temperature loop, which manipulates the setpoint of a heating or cooling coil or jacket. In this case, flow ratio gain is eliminated. Additionally, the process gain simplifies to about 1.0, giving Equation 1.1j, for the heating or cooling of a batch vessel when there is no reaction or feed. Equation 1.1j is Equation F.6h in Appendix F with a zero feed flow and zero heat of reaction. The simplification and linearization of the open loop gain is a significant advantage in addition to the compensation of cooling water or steam pressure or temperature changes by the coil or jacket loop before they appreciably affect vessel temperature.

For cascade control where an upper vessel temperature loop is manipulating the lower coil or jacket temperature loop, the open loop gain simplifies to the product of the setpoint and measurement scale gains for zero feed flow and zero heat of reaction, yielding:

$$K_o = K_{sp} * K_m \tag{1.1j}$$

Note that an important check as to the correctness of a calculation in terms of units and even expressions is that the open loop steady state gain is dimensionless.

For cascade control of vessel temperature to coil or jacket temperature, the open loop steady state gain can simplify to being linear and a function of temperature scale spans.

For integrating processes, the open loop self-regulating process (steady state) gain (K_o) is replaced with an open loop integrating process gain (K_i) with inverse time units (1/sec), which results from the process gain term units not cancelling out the flow time units as seen in Equation 1.2a through 1.2c for level. For level we can eliminate the ratio gain term defined by Equation 1.1d in the open loop gain (K_o) for a self-regulating process, giving an equation for the open loop integrating process gain (K_i) the distinction of no longer being dimensionless but having units of 1/sec.

$$K_i = K_v * K_p * K_m \tag{1.2a}$$

$$K_v = \frac{\Delta F_v}{\Delta \% CO} \tag{1.2b}$$

$$K_p = \frac{1}{A * \rho} \tag{1.2c}$$

$$K_m = \frac{\Delta \% PV}{\Delta L} \tag{1.2d}$$

One can convert an open loop self-regulating process (steady state) gain (K_o) for a process with a primary time constant (τ_o) much larger than the dead time (θ_o) to an open loop integrating process gain (K_i) by the use of Equation 1.3a. This equation was developed for near-integrator approximation, enabling the use of tuning rules for integrating processes to improve disturbance rejection. The equation can also save significant amounts of time in the identification of the process dynamics, since the largest ramp rate $Max(\Delta\%PV/\Delta t)$ in two to four dead times per Equation 1.3b is used to compute the integrating process gain, instead of waiting for the process to essentially reach a steady state (i.e., 98% response time) to identify the primary time constant and the open loop steady state gain. For a time constant that is 20 times the dead time (not uncommon for well-mixed liquid reactors), the identification time is 20 times faster.

$$K_i = \frac{K_o}{\tau_o} \tag{1.3a}$$

$$K_i = \frac{Max(\Delta\%PV/\Delta t)}{\Delta\%CO} \tag{1.3b}$$

The conversion of the open loop steady state gain and primary time constant of a self-regulating process to an open loop integrating process gain has many benefits in PID tuning and in the identification and analysis of process dynamics.

For integrating processes, the product of the controller gain (PID gain) (K_c) and the integral time (reset time) setting (T_i) must be greater than twice the inverse of the open loop integrating process gain (K_i) to prevent the start of slow rolling oscillations:

$$K_c * T_i > 2 * K_i \tag{1.3c}$$

Since most PID on integrating processes have a controller gain much less than the maximum allowed, the equation is reformulated to show the minimum integral time:

$$T_i > \frac{2}{K_c * K_i} \tag{1.3d}$$

The oscillations will decay slowly unless the following inequality is enforced:

$$T_i > \frac{1}{4 * K_c * K_i} \tag{1.3e}$$

Recent test results show that the above rules apply to near-integrating and runaway processes as well. As one goes from near-integrating to true integrating to runaway processes, the consequences of violating these rules get more severe in that the oscillations are larger and are slower to decay.

If the PID gain is much less than the maximum allowed, the reset time must be increased to prevent the start of slow nearly undamped oscillations for near-integrating, true integrating, and runaway processes.

Most of the tuning problems associated with level control, gas pressure control of large volumes and composition, pH, and temperature control of large well-mixed liquid volumes stems from a reset time that is orders of magnitude too small (e.g., reset time is 100 sec instead of 10000 sec). The reset time is too small because reset provides the type of response a human would provide (no sense of direction or anticipation, as noted in the Appendix B example). Gain settings are an order of magnitude smaller than the maximum allowed because gain settings above 10 are not described in the literature, so the operator is concerned about full scale PID output changes for setpoint changes. Simulations can show the value of higher gain settings. Setpoint rate limits can be added to the signal manipulating a valve, VSD, or secondary flow loop (i.e., effectively move suppression) to make these changes less disruptive to other loops if the PID has external reset feedback.

The reset time setting for level, gas pressure, and well-mixed liquid volume composition, pH, and temperature control is frequently orders of magnitude too small.

Nomenclature for PID Gain and Reset Time Calculations

A = cross-sectional area of vessel (m^2)
K_c = controller gain (dimensionless)
K_m = measurement gain (% / PV e.u.)
K_i = open loop integrating process gain (% / % / sec) (1 / sec)
K_o = open loop self-regulating process (steady state) gain (% / %) (dimensionless)
K_p = process gain (PV e.u.)
K_r = flow ratio gain often imbedded in process gain (1 / flow e.u.)
K_v = valve or variable speed drive gain (manipulated flow per % output)

$\Delta\%CO$ = change in controller output converted to % of controller output scale (%)
ΔF_v = change in valve or variable speed drive flow (flow e.u.)
F_p = process flow at current production rate (flow e.u.)
ΔL = change in level (m)
$\Delta\%PV$ = change in process variable converted to % of controller input scale (%)
PV_{max} = maximum process variable scale value (PV e.u.)
PV_{min} = minimum process variable scale value (PV e.u.)
ΔR = change in ratio of manipulated flow to process flow (dimensionless)
SP_{max} = maximum setpoint scale value (SP e.u.)
SP_{min} = minimum setpoint scale value (SP e.u.)
T_i = integral time (reset time) (sec)
τ_o = primary time constant (largest time constant in loop) (sec)
ρ = liquid density in vessel (kg/m^3)

1-2. Types of Process Dynamics

There are three fundamental types of process dynamic responses (self-regulating, integrating, and runaway) depending upon the sign of feedback within the process. The sign of the feedback is apparent from whether the response approaches a final value, continues to ramp, or accelerates in the latter portion of the open loop response. The ordinary differential equations for the process documented in Appendix F show the first principle origin of the process feedback sign and the terms that define the corresponding open loop response. When the degree of feedback within the process is small within the PID time frame due to a large process time constant, the process can be treated as if there is no feedback.

> ***Process dynamic responses depend upon the sign and degree of feedback within the process.***

1.2.1 Self-Regulating Processes

The most common type of process has internal negative feedback. These processes are self-regulating in that they will settle out at a steady state value as shown in Figure 1.2a for a given step change in PID controller output assuming there are no process disturbances. Liquid flow, liquid pressure, continuous concentration, and open loop stable continuous temperature processes have a self-regulating response.

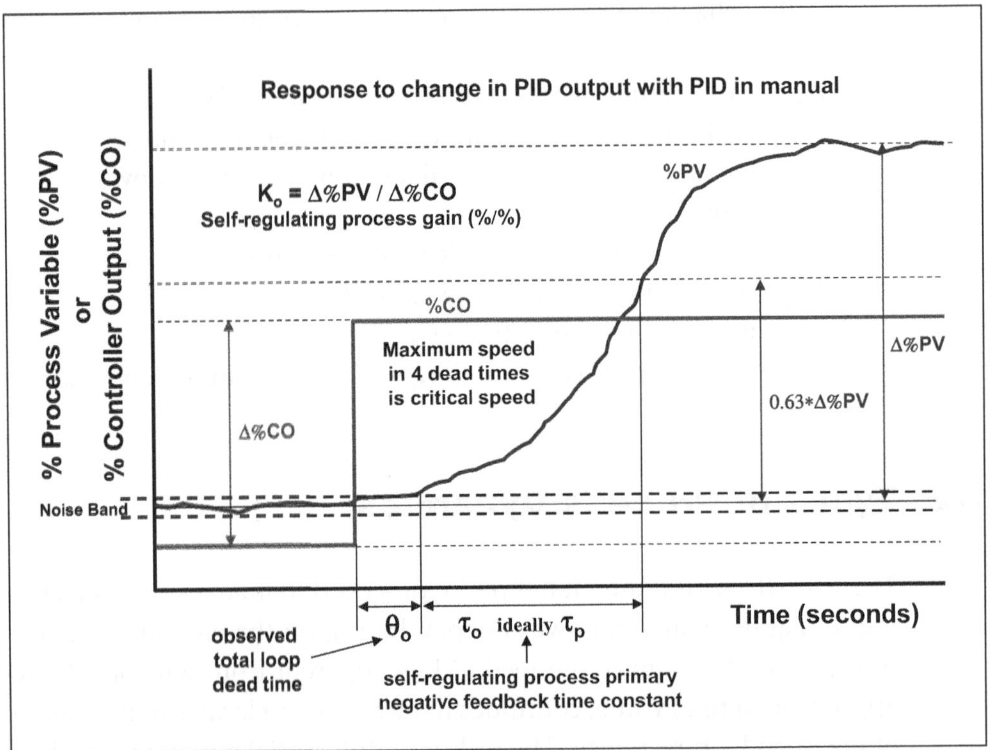

Figure 1.2a – Dynamics for Negative Feedback Process (Self-Regulating Process)

> *The most common type of process has internal negative feedback and is called self-regulating because the process will eventually settle out at final value for a given change in PID output.*

The primary time constant is the time to reach 63% of the final change in PV value (%ΔPV) after the PV starts to change for a given change in the controller output (%ΔCO). The primary time constant is really the largest time constant in the loop. Ideally, the largest time constant is in the process but for liquid flow, pressure, and some inline processes, the largest time constant originates in the sensor, transmitter, or PID filter. Some software can identify the second largest time constant, which is called the secondary time constant. This identification can help in the tuning of controllers that use derivative action because setting the rate time equal to the secondary time constant cancels out its detrimental effect in slowing down closed loop control (control when the PID is in AUTO, CAS, or RCAS modes). If the process response had no secondary time constant, the slope of the response would be at its maximum value immediately after the dead time. The secondary time constant creates a gradual increase (e.g., bend) in the rate of change of the process after the dead time. All processes have a secondary time constant. If this time constant is not identified, the secondary time constant is converted to an equivalent dead time to provide a first order plus dead time (FOPDT) model. If the secondary time constant is identified we have a second order plus dead time model.

If the primary time constant is less than ¼ the total loop dead time, we have a sub category of a self-regulating process called a dead time dominant process. The process response to a step change in a PID output is essentially a step change in the process variable. These processes are more vulnerable to process noise due to lack of the filtering action within the process. The tuning of such processes requires significantly more integral and less proportional action than all other processes. Derivative action is not used unless there are multiple time constants and no noise.

> ***A self-regulating process with the primary process time constant less than ¼ the dead time is termed dead time dominant. These processes benefit from less proportional action and usually no derivative action but more integral action.***

If the process time constant is greater than 4 times the total loop dead time, we have a sub category of self-regulating process called near-integrating because of the degree of negative feedback within the time frame of the PID response. The tuning for near-integrating processes is nearly the same as for true integrating processes, which have zero internal feedback.

The near-integrating process classification is based on the realization that a PID does most of its correction in the first four dead times during which the process has not appreciably decelerated. Concentration, pH, and temperature control of continuous unit operations with back mixed liquid volumes from boiling, reflux, recirculation, and agitation have a near-integrating response. Some examples are distillation columns, crystallizers, continuous stirred tank reactors (CSTR), evaporators and neutralizers.

A self-regulating process with a primary process time constant greater than 4 times the dead time is termed near-integrating. Many of the important control loops on liquid unit operations have a true or near-integrating response. These processes benefit from more proportional and derivative action and less integral action.

When the time constant is between ¼ and 4 times the dead time, we have a moderate self-regulating process. These primarily occur in the continuous gas unit operations in the front end of refineries and hydrocarbon processes. Most of the control literature on the design and implementation of PID, Model Predictive Control (MPC), or special control algorithms addresses moderate self-regulating processes.

Most of the control literature on PID control is based on a moderate self-regulating process where the primary process time constant is between ¼ and 4 times the dead time.

The open loop steady state gain is dimensionless and is the final change in percent PV for the step change percent change in PID output. Note that the changes in PV must be in percent of input and output scales because the PID algorithm works in percent and not engineering units.

1.2.2 Integrating Processes

The next most common type of process has zero internal feedback. These processes are integrating and will ramp continually off to an equipment or process limit as shown in Figure 1.2b for a given step change in the PID output. Level, gas pressure, and batch concentration, pH, and temperature processes have an integrating response.

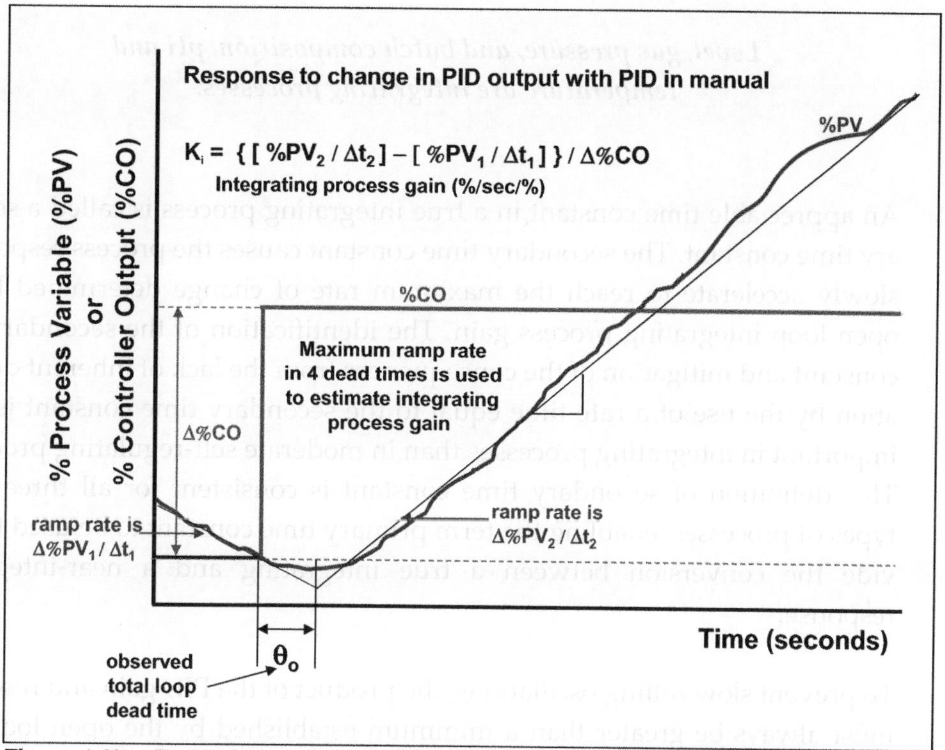

Figure 1.2b – Dynamics for Zero Feedback Process (Integrating Process)

The key characteristic of a zero feedback process is an open loop integrating process gain that is the change in ramp rate of the PV in percent per second ($\%\Delta PV_2$/sec - $\%\Delta PV_1$/sec) divided by the step change in percent controller output (%ΔCO), assuming there are no process disturbances. If the PID was in AUTO, CAS, or RCAS modes immediately before the test, the initial ramp rate is typically zero ($\%\Delta PV_1/sec = 0$). The ramp rate of most integrating processes is extremely slow (e.g., 0.000001 %/sec).

The open loop integrating process gain can be converted to a primary time constant and open loop steady state gain of a near-integrating process and vice versa via Equation 1.3a. The conversion is useful for gaining a better understanding and a set of tuning settings. In fact there are few really true integrating processes. If there were no equipment, process, or measurement limits nearly all integrating processes would reach a steady state. For example if the liquid level or gas pressure in a vessel could increase without bounds, the additional head or pressure would increase the liquid flow or gas flow out of the vessel to balance the increase in flow into the vessel.

Level, gas pressure, and batch composition, pH and temperature are integrating processes.

An appreciable time constant in a true integrating process is called a secondary time constant. The secondary time constant causes the process response to slowly accelerate to reach the maximum rate of change determined by the open loop integrating process gain. The identification of the secondary time constant and mitigation of the consequences from the lack of inherent deceleration by the use of a rate time equal to the secondary time constant is more important in integrating processes than in moderate self-regulating processes. This definition of secondary time constant is consistent for all three major types of processes, enabling the term primary time constant to be used to provide the conversion between a true integrating and a near-integrating response.

To prevent slow rolling oscillations, the product of the PID gain and reset time must always be greater than a minimum established by the open loop integrating process gain, as defined by Equation 1.3e. Most integrating processes have a gain setting much lower than is permitted by the process dynamics. Since people associate oscillations with too high of a gain and decrease the gain, they make the problem worse when the solution was to increase either the gain or the reset time. Most PID controllers on integrating processes that do not use tuning settings identified by software have too much integral action. For large liquid volumes the software identified setting for PID gain are uncomfortably large; in these cases the gain setting is consequently significantly reduced, thinking this makes the loop less likely to oscillate when in fact the loss of negative feedback control action is detrimental to a process without internal negative feedback. The solution is to increase the reset time by the same factor to keep the product of the PID gain and reset time the same as computed by the software. The first thing to try in most cases that the period of the oscillation is more than 10 times the dead time is to simply increase the reset time by a factor of 100 to 1000. Whereas decreasing or increasing the PID gain can cause slow or fast oscillations, respectively, increasing the reset time will always make the process more stable.

Processes without appreciable internal negative feedback have a window of allowable PID gains in which a gain too small and a gain too large will cause slow and fast oscillations, respectively.

PID controllers on integrating processes that are tuned manually generally have a reset time that is orders of magnitude too large, causing slow rolling oscillations.

Increasing the reset time always makes the process response smoother and more stable.

1.2.3 Runaway Processes

Processes with internal positive feedback fortunately do not occur very often but are associated with some extremely important unit operations. These processes have an open loop runaway response where the PV will continue to accelerate as shown in Figure 1.2c for a given step change in the PID output. These processes are called open loop unstable.

The most common example is an exothermic reactor where the heat of reaction increases with temperature and can exceed the cooling rate if the temperature gets high enough. A much less common example is a runaway speed condition when a large compressor goes into surge when an axial compressor has extremely low rotor inertia. The cell mass concentration control of a biological reactor can exhibit a positive feedback response in the exponential growth phase where an increase in cell concentration increases the cell growth rate. A runaway condition in such a reactor is not typically a concern because the acceleration in the cell mass concentration is so slow and is followed by a stationary phase where the acceleration stops due to the death rate of older cells equaling the growth rate. The pH response of a true strong acid and strong base neutralizer will dramatically accelerate as the pH approaches neutrality (the rate of change of pH increases by a factor of ten for each pH unit in the approach to neutrality). Fortunately most systems are moderated by slight concentrations of weak acids and weak bases, most notably carbonic acid associated by exposure to what might seem to be insignificant concentrations of carbon dioxide.

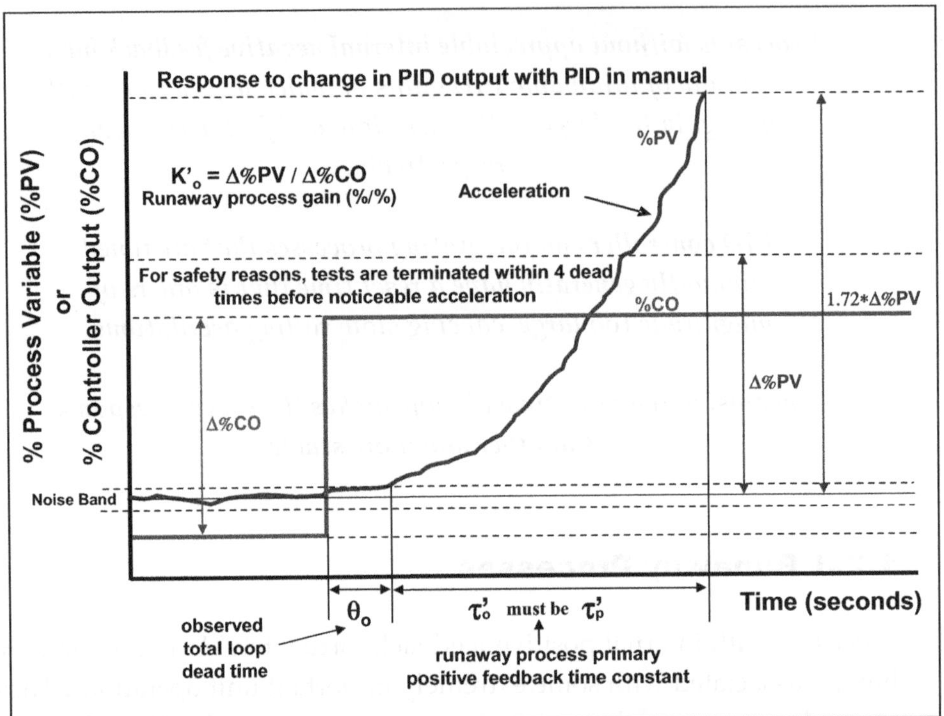

Figure 1.2c – Dynamics for Positive Feedback Process (Runaway Process)

> ***A runaway response can develop for temperature control when a chemical reactor heat of reaction exceeds the cooling rate or for biomass concentration control when a bioreactor cell growth rate exceeds the death rate.***

The primary process time constant for a process with positive feedback is the time for the PV to reach 172% of the value established by the open loop runaway process gain. This gain can be computed as shown in Appendix F but roughly corresponds to the point where a noticeable continual acceleration becomes significant. The identification of the positive feedback time constant and open loop gain by an open loop test is rarely done in practice because of the danger of a runaway. These loops often have to stay in AUTO mode at all times so that the PID controller provides through its control algorithm the negative feedback missing in the process that is needed for safe operation.

Runaway processes are tuned as integrating processes. The equations in Appendix F for positive feedback time constant and gain are used for control system design and analysis.

Runaway processes are identified as integrating processes. Even when a runaway process exists, tests are designed not to show the acceleration because such acceleration is unsafe. The tuning must be aggressive in that the process will run away if the PID gain is too small to provide enough proportional action. In the next section equations are presented to compute the maximum and minimum gain for stability as a function of dead time, the primary time constant and the secondary time constant. Proportional action is maximized to provide a more immediate reaction. Integral action is minimized (reset time is maximized) because the integral mode will not seek to add coolant until the PV has crossed setpoint leading to overshoot. Derivative action is more important than in other processes in order to help prevent excessive acceleration and to compensate for a secondary time constant.

Processes with positive feedback have a window of allowable PID gains where a PID gain too small and a PID gain too large can both cause a dangerous runaway condition.

1-3. Controller Tuning

There are over 400 tuning rules compiled in Aidan O'Dwyer's *Handbook of PI and PID Controller Tuner Rules (Ref 75)*. The author of each of these rules thinks their method is best. The diversity of tuning rules is primarily due to different types of process responses (e.g., dead time dominant self-regulating, moderate self-regulating, near-integrating, true integrating, and runaway processes), unknowns, nonlinearities, different objectives (e.g., setpoint response, load response, maximum variability absorption), and extenuating circumstances (e.g., interactions, inverse response, and resonance). The existence of runaway processes and the tuning guidelines have been directly addressed only by Bill Luyben in "Proportional-Derivative Control of Exothermic Reactors" and "Consider Reactor Control Lags" (*Ref 33 and 34*). In the absence of additional guidelines, runaway processes should be treated as integrating processes. The use of derivative action is very important to help react to acceleration and provide anticipation.

If the effective rate time becomes larger than the reset time, oscillations can break out even if the tuning settings would otherwise provide stable operation. The Series PID Form used in analog controllers and DCS systems has inherently prevented the effective rate time from becoming more than the effective reset time by an interaction factor. Some tuning rules developed in the 1960s used rate times approximately equal to the reset time. The use of these settings can cause severe oscillations in today's DCS. There is no protection in the ISA Standard PID Form predominantly used today to prevent the rate time equaling or exceeding the reset time. To ensure the rate time is well below the reset time, a high limit for the rate time that is ¼ the reset time is shown in this book.

There is no protection in the ISA Standard Form used today to prevent the effective rate time equaling or exceeding the effective reset time causing severe oscillations.

The user must also be extremely careful about the units of the tuning settings and PID output scale. Some controllers use proportional band instead of gain for the proportional mode and repeats per minute instead of seconds for the integral mode. These settings have inverse relationships that can lead to enormous errors in tuning and performance. The output scale of the modern DCS is in engineering units even though the algorithm works with percent signals and the default output limits and anti-reset windup limits are in percent. The analog and early DCS PID controllers had output scales in percent.

In migration projects, the tuning settings and output and anti-reset windup limits must be carefully converted based on the PID Form, tuning setting units, and the PID output scale.

Most of the control literature focuses on improving the response to a setpoint or a disturbance on the process output for moderate self-regulating processes. Gamesmanship tends to rule and algorithms and tuning methods are touted based on improvements without a realization of the diversity of conditions and demands in actual industrial installations.

Lambda tuning rules are presented here because of their flexibility in dealing with different processes, nonlinearities, objectives, and extenuating circumstances. Modifications are presented to the normal lambda tuning rules to enable lambda tuning to give a similar load disturbance rejection capability as the Short Cut Method when this is the criterion without interfering with the flexibility of lambda tuning.

A Short Cut Method (SCM) developed by the author is used to provide a benchmark of aggressive tuning settings that will minimize the peak and integrated error from load disturbances (process input disturbances) without causing an oscillatory response if the dynamics are perfectly known and constant. The Short Cut Method first estimates the ultimate period, which has many diagnostic uses beyond the computation of tuning settings.

The estimation or identification of the ultimate period has many diagnostic uses.

Consider the situation in which there is no oscillatory disturbance. If the loop oscillation period is less than ¾ the ultimate period, the rate time is too large. If the oscillation period is close to the ultimate period, the PID gain is too large. If the oscillation period is greater than twice the ultimate period, the PID reset time is too small. If the oscillation period is greater than four times the ultimate period and the process is near-integrating, true integrating, or runaway, the product of the PID gain and reset time is too small.

Now consider an oscillatory disturbance. If the period of an oscillatory disturbance is between ¼ and 4 times the ultimate period, the PID does more harm than good due to amplification from resonance from feedback control. The oscillatory disturbance period and amplitude is observable when the PID is in manual. For cascade control, the ultimate period of the secondary loop should be 5 times faster than the ultimate period of the primary loop. For minimizing interactions the same rule applies as to the relative size of the ultimate period.

Other tuning methods can be used as long as the advocate uses good software to compute the settings and the window of allowable controller gains for integrating and runaway processes is not violated. More important than the tuning rules is the actual tuning of the loop with good software, including any change in settings by an experienced practitioner to address unknowns, nonlinearities, varying objectives, and extenuating circumstances. Consultants often tout their tuning rules when in actuality the process of intelligently tuning a loop is more important than the rules used. PID tuning totally set manually, based on intuition, is in most cases messed up.

Consultants have such pride and time invested in their tuning rules that the improvement in a loop is often more attributed to the rules than to the software, which gets settings based on identified dynamics and the intelligence used to modify the settings to deal with the conditions and meet the demands of industrial applications.

The use of lambda rather than lambda factors offers many important advantages. Consultants who use lambda tuning for industrial processes do not enter a lambda factor but an actual lambda that is a closed loop time constant for self-regulating processes and an arrest time for integrating processes. Lambda typically ranges from one dead time to 3 dead times to meet application requirements. Limits are imposed on the reset and rate times to help maximize load rejection. Finally processes with a time constant to dead time ratio greater than four are treated as near-integrating processes, and lambda tuning rules for integrating processes are used.

The use of lambda rather than lambda factors is essential. Also, the addition of simple limits on reset and rate time and recognizing near-integrating processes enables lambda tuning rules to maximize disturbance rejection without a loss of flexibility.

1.3.1 Lambda Tuning for Self-Regulating Processes

The lambda tuning method for self-regulating processes is advocated by the author for use when the open loop time constant is less than 4 times the total loop dead time. For self-regulating processes that are not near-integrating ($\tau_o < 4 * \theta_o$) we have the following series of equations.

The reset time is set equal to the open loop time constant (Equation 1.4a). The reset time steadily decreases as the ratio of the time constant to dead time decreases from its maximum of 4 dead times if this tuning method is used when the open loop time constant is less than 4 times the total loop dead time. For systems with an open loop time constant much less than the dead time (dead time dominant), the PID action becomes essentially integral only due to a steady decrease in the controller gain and the reset time. This type of control helps deal with the abrupt, almost step changes and noise in a dead time dominant system. Thus, the PID controller takes over the role of providing a gradual and smoothing response that is missing in the process. The best scenario is to have a large process time constant to fulfill this role but for plug flow and sheet line processes, this is not possible.

$$T_i = \tau_o \quad (1.4a)$$

A low limit is added to traditional equations to prevent a reset time smaller than 0.4 times the dead time when the time constant is much smaller than the dead time (Equation 1.4b). This low limit provides more proportional mode action (Equation 1.4c). Keeping the PID gain from approaching zero gives a more immediate response to disturbances and setpoint changes and helps get the PID output get through valve or VSD deadband and resolution limits.

$$T_i = Max[0.4 * \theta_o, \tau_o] \quad (1.4b)$$

$$K_c = \frac{T_i}{K_o * (\lambda_f * \tau_o + \theta_o)} \quad (1.4c)$$

For dead time dominant processes, a low limit of 0.4 times the dead time on the reset time provides some proportional action unless nearly integral-only control is desired.

In industrial applications the lambda factor multiplication of the open loop time constant is replaced with lambda, the closed loop time constant ($\lambda = \lambda_f * \tau_o$), which is the time after the dead time to reach 63% of a setpoint change. The use of lambda rather than a lambda factor provides the recognition that lambda is set relative to the dead time to minimize variability in the PID input or output and to provide the gain margin needed to deal with extenuating circumstances (e.g., interaction, inverse response, and resonance).

$$K_c = \frac{T_i}{K_o * (\lambda + \theta_o)} \tag{1.4d}$$

For maximum unmeasured disturbance rejection, a lambda equal to the dead time is used giving the Equation 1.4e for the PID gain setting when the reset time is not limited. For maximum disturbance rejection of self-regulating process, a lambda equal to 0.6 times the dead time is used giving nearly the same PID gain as the original Ziegler Nichols reaction curve method defined in the landmark paper "Process Lags in Automatic Control Circuits" (*Ref 101*).

$$K_c = 0.5 * \frac{\tau_o}{K_o * \theta_o} \tag{1.4e}$$

A rate setting is not normally included as part of the lambda tuning for self-regulating processes. If the primary time constant is greater than ½ the dead time, rate action may be beneficial. If a secondary time constant (the next largest time constant) can be identified, the rate time is set equal to the secondary time constant. The rate time should be larger than ½ the dead time but not be greater than ¼ the reset time for an ISA Standard Form. The derivative mode must have a built-in filter whose time constant is a fraction of the rate time. The fraction is the inverse of the alpha setting (e.g., 8).

> ***When derivative action is beneficial the rate time should be the secondary time constant or ½ the dead time whichever is largest but not be greater than ¼ the reset time for an ISA Standard Form. The derivative mode must have a built-in filter whose time constant is a fraction of the rate time. The fraction is the inverse of the alpha setting (e.g., 8).***

If the primary time constant is greater than ½ the dead time ($\tau_o > 0.5 * \theta_o$):

$$T_d = Min[0.25 * T_i, Max(0.5 * \theta_o, \tau_s)] \qquad (1.4f)$$

For maximum disturbance rejection of self-regulating process, a lambda equal to 0.6 times the dead time is used.

1.3.2 Lambda Tuning for Integrating Processes

The lambda tuning method for integrating processes is advocated by the author for use on self-regulating processes when the open loop time constant is greater than 4 times the total loop dead time (near-integrating processes) in addition to use on true integrating and runaway processes to provide a faster return to setpoint for load upsets.

Near-integrating, true integrating, and runaway processes should use integrating process tuning rules with a low limit of 4 dead times for the reset time when derivative action is not used.

For PID, the reset time is twice the arrest time plus the dead time.

$$T_i = 2 * \lambda + \theta_o \qquad (1.5a)$$

For PI, a low limit is added so that the reset time is not less than 4 dead times.

$$T_i = Max[4 * \theta_o, 2 * \lambda + \theta_o] \qquad (1.5b)$$

The original equation for controller gain:

$$K_c = \frac{T_i}{K_i * \left(\frac{\lambda_f}{K_i} + \theta_o\right)^2} \qquad (1.5c)$$

In industrial applications the lambda factor multiplication of the inverse of the open loop integrating process gain is replaced with lambda, the closed loop arrest time ($\lambda = \lambda_f / K_i$), which is the time to stop an excursion for an unmeasured disturbance (time to peak error). The use of lambda rather than lambda factor provides the recognition that lambda is set as a multiple of the dead time to minimize variability in either the PID input or output and to provide the gain margin needed to deal with extenuating circumstances (e.g., interaction, inverse response, and resonance) or is set per Section 1.3.7 to provide the maximum absorption of variability.

The numerator uses the unlimited reset time.

$$K_c = \frac{2 * \lambda + \theta_o}{K_i * (\lambda + \theta_o)^2} \tag{1.5d}$$

Near-integrating, true integrating, and runaway processes should use integrating process tuning rules with a low limit of 4 dead times for the reset time. The unlimited reset time is used in the numerator of the lambda calculation for the PID gain.

For maximum unmeasured disturbance rejection with some robustness, a lambda equal to the dead time is used:

$$K_c = 0.75 * \frac{1}{K_i * \theta_o} \tag{1.5e}$$

A lambda equal to 3 dead times minimizes the consequences of nonlinearities, inverse response, and resonance.

Equation 1.5f for the rate time when the primary time constant is more than ½ the dead time in a self-regulating process is used for integrating processes:

$$T_d = Min[0.25 * T_i, Max(0.5 * \theta_o, \tau_s)] \tag{1.5f}$$

1.3.3 Lambda Tuning for Runaway Processes

For runaway processes, the lambda tuning for integrating processes is used. The concept of the positive feedback process time constant and acceleration in the open loop response is more for process understanding than for tuning. For safety reasons, open loop tests are not done on highly exothermic reactors. Instead, a closed loop test is done for a step in the PID output from a setpoint change with a PID Structure "PI on error and D on PV" and no setpoint filter or lead-lag. The dead time, secondary time constant, and open loop integrating process gain are identified in the initial response before integral action has had much of an effect. A more accurate identification can be done by temporarily increasing the reset time by an order of magnitude or more.

1.3.4 Short Cut Method Tuning for Self-Regulating Processes

The ultimate period can be estimated from the primary time constant and the total loop dead time based on Bode plot results (Equation 1.6a). The PID gain is limited to being greater than 20% of the inverse of the open loop self-regulating process gain to prevent too small of a PID gain for dead time dominant loops. The settings computed here provide the most aggressive response to a load upset. In practice, the gain is reduced to provide a smoother and more robust response.

If the primary time constant (τ_o) is very large compared to the dead time (θ_o), the expression for the ratio of the time constant to the 63% response time (dead time plus time constant) approaches 1 giving an ultimate period of 4 dead times. If the time constant is very small relative to the dead time, the expression approaches zero giving an ultimate period of 2 dead times:

$$T_u = 2 * \left[1 + \left(\frac{\tau_o}{\tau_o + \theta_o}\right)^{0.65}\right] * \theta_o \qquad (1.6a)$$

> ***The ultimate period for self-regulating processes decreases from 4 dead times to 2 dead times as the ratio of the primary time constant to total loop dead time approaches zero.***

For an ISA Standard Form proportional-integral (PI) controller:

$$K_c = Max\left(0.2/K_o, 0.6 * \frac{\tau_o}{K_o * \theta_o}\right) \qquad (1.6b)$$

$$T_i = 0.8 * \frac{T_u}{Min\left(4,\ 10 * \left(\frac{4 * \theta_o}{T_u} - 1\right)^2 + 1\right)} \quad (1.6c)$$

For an ISA Standard Form proportional-integral-derivative (PID) controller:

$$K_c = Max\left(0.2/K_o,\ 0.8 * \frac{\tau_o}{K_o * \theta_o}\right) \quad (1.6d)$$

$$T_i = 0.6 * \frac{T_u}{Min\left(4,\ 10 * \left(\frac{4 * \theta_o}{T_u} - 1\right)^2 + 1\right)} \quad (1.6e)$$

The optimum reset time for self-regulating processes decreases from 4 dead times to about ½ the dead time as the ratio of the primary time constant to total loop dead time approaches zero.

The following expression automatically reduces the rate time to being simply the secondary time constant as loop becomes more dead time dominant indicated by the reset time approaching ½ the dead time. If the secondary time constant is not identified, it by definition is included in the dead time and the rate time becomes totally dependent upon the dead time and reset time.

$$T_d = Min[0.25 * T_i,\ 0.8 * Max[0.0,\ 0.25 * (T_i - 0.5 * \theta_o) + \tau_s]] \quad (1.6f)$$

1.3.5 Short Cut Method Tuning for Integrating Processes

The ultimate period can be estimated from the secondary time constant and total loop dead time based on Nyquist plot results. The PID gain is limited to being less than the valve stick-slip (S_v) divided by the difference between the measurement noise (N_m) and the measurement threshold sensitivity (S_m) to prevent fluctuations in the PID output from exceeding the stick-slip causing excessive packing wear and high frequency disturbances. The divisor is limited to a 16 bit analog/digital (A/D) convertor resolution of 0.003%. In older DCS with a 12 bit A/D, a resolution of 0.05% should be used in divisor.

If the secondary time constant (τ_s) is very small compared to the dead time (θ_o), the expression for the ratio of the time constant to the dead time approaches zero giving an ultimate period of 4 dead times. If the time constant is large relative to the dead time, the expression gets large giving an ultimate period that is a large multiple of the dead time (e.g., 10 dead times):

$$T_u = 4 * \left[1 + \left(\frac{\tau_s}{\theta_o}\right)^{0.65}\right] * \theta_o \tag{1.7a}$$

The ultimate period for integrating processes increases from 4 dead times as the ratio of the secondary time constant to total loop dead time increases.

For an ISA Standard Form proportional-integral (PI) controller:

$$K_c = Min\left(\frac{S_v}{\max[(N_m - S_m), 0.003]}, 0.6 * \frac{1}{K_i * \theta_o}\right) \tag{1.7b}$$

$$T_i = 0.8 * T_u \tag{1.7c}$$

For an ISA Standard Form proportional-integral-derivative (PID) controller:

$$K_c = Min\left(\frac{S_v}{\max[(N_m - S_m), 0.003]}, 0.8 * \frac{1}{K_i * \theta_o}\right) \tag{1.7d}$$

$$T_i = 0.6 * T_u \tag{1.7e}$$

$$T_d = Min[0.25 * T_i, 0.8 * Max[0.0, 0.25 * (T_i - 0.5 * \theta_o) + \tau_s]] \tag{1.7f}$$

1.3.6 Short Cut Method Tuning for Runaway Processes

The ultimate period can be estimated from the positive feedback time constant defined by Equations F.2a and F.2b in Appendix F, while the secondary time constant and total loop dead time are based on Nyquist plot results (Equations 1.8a, 1.8b and 1.8c). The PID gain is limited to being greater than twice the inverse of the open loop runaway process gain (K_o) to prevent the process from going unstable due to insufficient feedback action. PI tuning is not offered since the omission of derivative action is not advisable.

$$T_u = 4 * \left[1 + \left(\frac{N}{D}\right)^{0.65}\right] * \theta_o \qquad (1.8a)$$

$$N = (\tau_p' + \tau_s) * (\tau_p' * \tau_s) \qquad (1.8b)$$

$$D = (\tau_p' - \tau_s) * (\tau_p' - \theta_o) * \theta_o \qquad (1.8c)$$

If the secondary time constant (τ_s) is very small compared to the dead time (θ_o) and both of these parameters are small compared to the positive feedback time constant, the expression for the ratio of the time constant to the dead time approaches zero giving an ultimate period of 4 dead times. For the converse situation the expression gets very large giving an ultimate period that is a large multiple of the dead time.

> ***The ultimate period for runaway processes increases from 4 dead times as the ratio of the secondary time constant to total loop dead time increases and the ratio of the secondary time constant or total loop dead time to the positive feedback time constant increases.***

For an ISA Standard Form proportional-integral-derivative (PID) controller:

$$K_c = Max\left(2.0/K'_o,\ 0.8 * \frac{\tau_p'}{K_o' * \theta_o}\right) \qquad (1.8d)$$

$$T_i = 0.6 * T_u \tag{1.8e}$$

$$T_d = Min[0.25 * T_i, 0.8 * Max[0.0, 0.25 * (T_i - 0.5 * \theta_o) + \tau_s]] \tag{1.8f}$$

For temperature control of exothermic reactors, the open loop runaway process gain is the product of the jacket or coil temperature gain, the positive feedback process gain from Equations F.2a and F.2b in Appendix F, and the temperature measurement gain:

$$K_o' = K_j * K_p' * K_m \tag{1.8g}$$

1.3.7 Maximizing Absorption of Variability Tuning for Surge Tank Level

When the absorption of variability must be maximized, lambda is chosen to be as large as possible. The most common occurrence is a level control application where the transfer of flow changes coming into the volume to flow changes going out of the volume is minimized. The incoming flow changes are absorbed as much as possible by allowing the level to change within operating limits, such as low and high alarm points.

For using the full available volume of a surge tank, a full scale residence time (t_{fsr}) is computed as the maximum volume ($A * \rho * \Delta L_{max}$), for the maximum span of the level measurement divided by the maximum flow capacity of the valve or VSD (ΔLF_{max}). This full scale residence time term is not to be confused with the actual residence time (operating volume divided by operating flow rate) used in process calculations in chemical engineering and throughout the rest of this book; actual residence time is particularly important for computing back mixed time constants or plug flow transportation delays and the time for reaction conversion. The following equations were developed based on the dead time being negligible, which is a reasonable assumption for level:

$$t_{fsr} = \frac{A * \rho * \Delta L_{max}}{\Delta F_{max}} \tag{1.9a}$$

$$K_c = 2 * \frac{\Delta\%CO_{max}}{\Delta\%PV_{max}} \tag{1.9b}$$

$$T_i = 2 * t_{fsr} * \frac{\Delta\%PV_{max}}{\Delta\%CO_{max}} = 2 * t_{fsr} * \frac{2}{K_c} = t_{fsr} * \frac{4}{K_c} \quad (1.9c)$$

For surge tank level control, tuning settings can be computed to maximize the absorption of variability that will minimize the changes in the manipulated discharge flow by using up all of the available volume without violating the window of allowable controller gains.

Nomenclature for Tuning Calculations

A = cross-sectional area of surge tank (m^2)
K_c = controller gain (dimensionless)
K_i = open loop integrating process gain (1/sec)
K_o = open loop self-regulating process (steady state) gain (%/%) (dimensionless)
K_o' = open loop runaway process gain (dimensionless)
K_j = jacket gain (jacket temperature scale span / 100%) (°C/%)
K_p' = reactor positive feedback process gain per Appendix F (°C reactor/°C jacket)
K_m = measurement gain (100% / reactor temperature scale span) (%/°C)
t_{fsr} = full scale residence time (sec)
T_i = controller integral time (reset time) (sec)
T_d = controller derivative time (rate time) (sec)
T_u = ultimate period (sec)
$\Delta\%CO_{max}$ = maximum available change in controller output (%)
$\Delta\%PV_{max}$ = maximum allowable change in process variable (%)
$\%SP$= operating point setpoint (%)
ΔF_{max} = maximum change in valve or variable speed drive flow (e.g., flow span) (m/sec)
ΔL_{max} = maximum change in level (e.g., level span) (m)
λ = lambda for closed loop time constant or arrest time (sec)
θ_o = total loop dead time (sec)
τ_o = primary time constant (open loop time constant - largest time constant in loop) (sec)
τ_s = secondary time constant (second largest time constant in loop) (sec)

τ_p' = positive feedback process time constant (largest time constant in loop) (sec)

ρ = liquid density in surge tank (kg/m^3)

1.3.8 Window of Allowable Controller Gains

The existence of a maximum PID gain is intuitive and extensively discussed or at least implied in control theory literature and courses. A PID gain just 20% higher than that computed by the Short Cut Method (SCM) can cause the start of an oscillatory response with a period slightly larger than the ultimate period estimated in the Short Cut Method. Alternately, just a 20% increase in dead time or open loop gain can correspondingly trigger these same oscillations. An increase of 100% can cause oscillations to grow in amplitude with a period close to the ultimate period. Thus, the maximum controller gain for stability can be considered to be approximately twice the Short Cut Method PID gain assuming that all of the dynamics are known and constant and there are no extenuating circumstances (e.g., interactions, inverse response, or resonance).

Rarely discussed in the literature is the existence of a low PID gain limit for near-integrating and true integrating processes when the integral mode is used and in runaway processes regardless of the modes used. In all of these cases, the loop depends upon gain action to provide negative feedback missing in the process dynamics.

For near-integrating, true integrating, and runaway processes, there is a window of allowable controller gains. Too high of a PID gain causes oscillations with a period close to the ultimate period and a growing amplitude (classic instability). Too low of a PID gain causes slow rolling oscillations with a period 10 times the ultimate period and a gradually decaying amplitude.

For near-integrating, true integrating, and runaway processes the minimum allowable PID gain to prevent slow rolling oscillations decreases as the reset time decreases and the open loop integrating process gain increases. The oscillations are extremely slow (e.g., 10 times the ultimate period). While the oscillations never grow, they become nearly undamped (the amplitude of each successive oscillation decays less than 20%) when the following inequality is violated:

$$K_{cmin} > \frac{1}{4 * T_i * K_i} \tag{1.10a}$$

For near-integrating, true integrating processes, and runaway processes the minimum allowable PID gain to prevent slow rolling oscillations decreases as the reset time decreases and open loop integrating process gain increases.

For runaway processes, the consequences of a low PID gain are more severe and potentially unsafe in that a reaction can go unstable from negative feedback action via the proportional mode being insufficient to counteract the acceleration from positive feedback in the process. The minimum PID gain per Equation 1.9b must be greater than the inverse of the open loop runaway process gain defined by Equation 1.7g for temperature control of an exothermic reactor as the product of the jacket temperature gain, the positive feedback process gain per Appendix F, and the temperature measurement gain.

$$K_{cmin} > \frac{1}{K_o'} \tag{1.10b}$$

For runaway processes, an unsafe condition develops when the PID gain is less than the inverse of the open loop runaway process gain.

Luyben developed equations in "Proportional-Derivative Control of Exothermic Reactors" and "Consider Reactor Control Lags" (*Ref 33 and 34*) to predict the ratio of the maximum to minimum controller gain to show how the window narrows as the heat transfer surface and temperature sensor time constant increase relative to other dynamics. Luyben recommends a ratio greater than 20 to ensure that unknowns and nonlinearities or less than optimal tuning do not cause instability.

$$R_c = 1 + (1 - R_m - R_h) * \left[\frac{R_m + R_h}{R_m * R_h} - 1 \right] \tag{1.11}$$

For temperature control of exothermic reactors that can runaway, the window of allowable controller gains close as the sum of the ratios of the measurement or heat transfer time constant to the positive feedback time constant approach one.

Nomenclature for Runaway Process Minimum and Maximum PID gain:

K_c = controller gain (dimensionless)

K_i = open loop integrating process gain (1/sec)

K_o = open loop runaway process gain (dimensionless)

K_{cmin} = minimum controller gain (dimensionless)

R_c = ratio of minimum to maximum controller gain (dimensionless)

R_h = ratio of heat removal to reactor positive feedback time constant (dimensionless)

R_m = ratio of measurement to reactor positive feedback time constant (dimensionless)

T_i = controller integral time (reset time) (sec)

1-4. Loop Performance

The primary metrics for load disturbances (disturbances at the process input) are peak error and integrated error (accumulated error). Figure 1.3 shows that the minimum peak error obtainable from feedback control for the most aggressive tuning is reached in one loop dead time. The minimum integrated error is the area of two right triangles placed back to back. A smaller PID gain increases the peak error. A slower integral action (larger reset time) most notably slows the return to setpoint.

Internal Model Control and lambda tuning methods have been criticized for too low of a PID gain and too large of a reset time for loops with a large primary process time constant that resulted from an original emphasis on setpoint response. The use of a lambda arrest time equal to 0.6 times the dead time and a switch from self-regulating to integrating process tuning for near-integrating processes with a PI low limit for reset time of 4 dead times and PID rate time of ½ the dead time results in tuning settings that is close to the optimum eliminating this criticism. Appendix C extensively discusses the relationships between these metrics and loop dynamics and PID tuning.

For the rejection of load disturbances, the metrics are peak error and integrated error.

1.4.1 Load Disturbance Accumulated Error

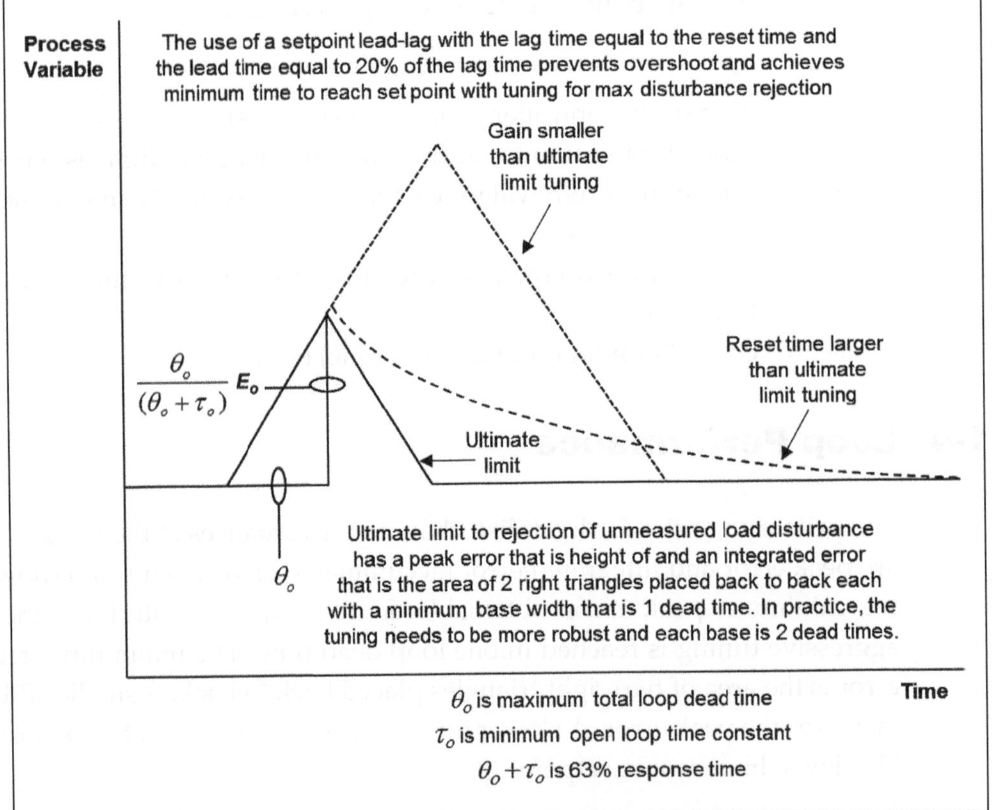

Figure 1.3 – Peak and Integrated Error Metrics for Load Response

Accumulated error is the totalized deviation of the controlled variable from setpoint. For a composition control loop (e.g., a distillation column temperature loop or a pH control loop), the accumulated error multiplied by the average product flow provides a measure of the total amount of product that deviates from the optimum product purity specification. For a flow control loop (e.g., a reactor feed loop or a furnace fuel loop), the accumulated error provides a measure of the total amount of feed that deviates from the stoichiometric ratio specification that may be important not only for product purity, but also for safety (if the excess feed is flammable, explosive, exothermically reactive, or toxic) or emission control (if the excess feed or associated by-products are environmentally restricted). If the controlled variable is a utility flow (e.g., steam flow or cooling water flow), the accumulated error is representative of energy usage in excess of the setpoint.

The accumulated error is the integrated error where the positive and negative errors are canceled by the volume of the system to provide a net accumulated positive or negative error. Integrated absolute error (IAE) is equal to the accumulated error for an overdamped or critically damped (non-oscillatory) response. Integrated squared error (ISE) can be approximated by a combination of accumulated error and peak error. The additional value of calculating the IAE and ISE errors is usually not worth the additional effort by the design or maintenance engineer for loop performance analysis. The exception occurs for split ranged flows where oscillations across the split range point can cause wasted energy resulting from alternating between cooling and heating and wasted reagent resulting from alternating between acid and base addition. However, a small measured accumulated error does not necessarily indicate a well-tuned stable loop. A small measured accumulated error can result from a loop that is marginally stable since the positive and negative errors will cancel for sustained oscillations. For a limit cycle, the accumulated error will generally steadily increase with time because the oscillation is rarely centered on the setpoint. The accumulated error can be accurately calculated by the use of a relatively simple algebraic equation if the oscillations are decaying in amplitude (Equation 1.15).

The accumulated error of the process variable with respect to setpoint provides a measure of the total quantity of reactants deviating from stoichiometric ratio, off-spec material produced, and the total quantity of excess energy, raw material, and reagent used.

The equations presented below for a self-regulating process can also be used for integrating and runaway processes. The open loop error is the ramp rate if the controller was in manual multiplied by the closed loop arrest time (lambda for integrating process tuning rules), which is a multiple of the dead time.

Equation 1.12 shows how the accumulated (integrated) error for a closed loop can be calculated for a given size load disturbance from the PID gain, reset setting, execution time, and filter time (*Shinskey, 1979*). The measurement gain (100% divided by measurement span) is included to get the error in percent of PID scale. The disturbance size is expressed as the percent change in PID output required (Δ%CO). For dead time compensators and the enhanced PID for wireless, the reset time can be decreased to the point where these equations will predict too low of an integrated error. The result of these equations cannot be taken to be less than the ultimate limit to the error as exemplified by Equation C.3 (Appendix C) and is determined by the amount of actual dead time in the loop.

$$E_i = \frac{T_i + \Delta t_x + \tau_f}{K_m * K_c} * \Delta\%CO \tag{1.12}$$

A more useful formulation of the equation uses the size of the open loop error (the error if the PID controller was in manual) from the disturbance. The change in valve or variable speed drive (VSD) flow, which is the change in controller output (Δ%CO) multiplied by the valve or VSD gain (K_v), must balance out the change in the disturbance variable (ΔDV) multiplied by the disturbance gain (K_d). The disturbance gain converts the load upset to the required change in manipulated flow:

$$\Delta\%CO * K_v = K_d * \Delta DV \tag{1.13}$$

The load multiplied by the gains gives the open loop error in engineering units:

$$E_o = K_d * K_r * K_p * \Delta DV \tag{1.14a}$$

Solving Equation 1.14a for ΔDV and substituting for ΔDV in Equation 1.13 cancels out K_d:

$$E_o = K_v * K_r * K_p * \Delta\%CO \quad (1.14b)$$

Solving Equation 1.14b for $\Delta\%CO$ and substituting the result into Equation 1.12 gives the integrated error in terms of the open loop error.

$$E_i = \frac{T_i + \Delta t_x + \tau_f}{K_c * K_v * K_r * K_p * K_m} * E_o \quad (1.14c)$$

$$K_o = K_v * K_r * K_p * K_m \quad (1.14d)$$

$$E_i = \frac{T_i + \Delta t_x + \tau_f}{K_c * K_o} * E_o \quad (1.15)$$

The accumulated error is proportional to the reset time and inversely proportional to the product of the PID gain and open loop self-regulating process gain.

The effect of the PID execution time or signal filter time on the accumulated error is negligible if the sum of these times is less than 10% of the reset time.

Nomenclature for Accumulated Error Metric:

E_i = integrated error from load disturbance (e.u. * sec)

E_o = open loop error (load disturbance error with controller in manual) (e.u.)

$\Delta\%CO$ = change in controller output converted to % of PID output scale (%)

ΔDV = change in disturbance variable (load) (e.u.)

K_c = controller gain (dimensionless)

K_d = disturbance gain (flow/e.u.)

K_m = measurement gain (%/e.u.)

K_o = open loop self-regulating process gain (%/%) (dimensionless)

K_p = process gain (e.u.)

K_r = flow ratio gain (often imbedded in process gain) (1/flow)

K_v = valve or variable speed drive gain (flow/%)
T_i = integral time (reset time) (sec)
τ_f = signal filter time constant (sec)
Δt_x = controller execution time (sec)

Equation 1.15 for the accumulated error leads to four conclusions that are important enough to emphasize. First, if the disturbances are nearly zero in magnitude ($E_o \approx 0.0$), even the most difficult loop will perform excellently. Thus before one can decide whether a difficult loop justifies the additional expense of special equipment, special instruments, or advanced control algorithms, knowledge of the size of the disturbance is necessary.

Second, if the controller was tuned with too small a gain (too large a proportional band) or too large a reset time (too slow a reset action) an easy loop will perform as poorly as a more difficult loop that required the higher mode settings. Any special efforts or expense during design to reduce dead time or increase a primary time constant will be wasted if overly conservative controller tuning is used in the field.

Third, if the PID execution time or filter time is large compared to the reset time, the accumulated error could be too large for even the best tuning.

Fourth, if the process gain is increased, the open loop error, and hence the accumulated error, increases per Equation 1.15 for a given disturbance. To recognize this relationship you need to realize that the K_p in the denominator is canceled by the K_p in the expression for PID gain K_c since K_c is inversely proportional to K_p. It is therefore important that the process engineer and control engineer review the effect of equipment design and operating conditions on the process gain at an early stage of the project (preferably before the project appropriation request is made final).

In applications where the measurement is an inference of a process condition (e.g., temperature is an inferential measurement of composition), a high process gain (high process sensitivity) provides better recognition of the disturbance ensuring a small change in the process condition is above the threshold sensitivity limit of the sensor and resolution limit of the PID. The resolution limit of wide range thermocouple input cards in the 1980s vintage DCS was about 0.25 degrees Centigrade, causing steps in the PID measurement that prevented the use of high PID gain and rate time settings. An increase in PID execution time and an increase in PID filter time settings helped to smooth the steps and increase the signal to noise ratio but the accumulated error per Equation 1.15 was adversely affected. In a modern DCS the resolution of the input cards is better than 0.02 degrees Centigrade, which exceeds the capability of the sensor.

> ***A PID controller tuned sluggishly will perform as badly as a loop with much more dead time. Any special efforts or expense during design to reduce dead time or increase a primary time constant will be wasted if overly conservative controller tuning is used in the field.***

1.4.2 Load Disturbance Peak Error

Peak error is the maximum deviation of the controlled variable from set point. For a reactor temperature or reactor pH loop, the maximum deviation might have to be limited to prevent the start of an undesirable secondary reaction. For a pressure control loop, the maximum deviation might have to be limited to prevent actuation of relief valves or vacuum breakers. For a Claus furnace combustion control loop, the maximum emission level of sulfur dioxide must meet environmental regulations (a Claus furnace produces sulfur from sulfur dioxide waste gas). For a waste treatment neutralizer, the pH must always be between low and high environmental limits to prevent a recordable incident. The peak error is equal to approximately 1.5 times the steady-state offset for a proportional-only controller per Equation 1.16 developed by Peter Harriott in the book *Process Control* (*Ref* 24).

> ***Limiting the peak error is important for preventing the triggering of undesirable reactions, high temperature and high pressure shutdowns, environmental violations (e.g., sulfur dioxide and Resource Conservation and Recovery Act pH), and the blowing of relief valves and rupture discs.***

For proportional-only (P-only) controller:

$$E_x = \frac{1.5}{K_c * K_o + 1.0} * E_o \tag{1.16}$$

The peak error is reached in about one dead time. The integral mode contribution is the number of repeats of the proportional mode contribution until the peak error is reached giving Equation 1.17. For processes with a small dead time to time constant ratio, the integral time is much larger than the dead time and the peak error is mostly affected by the proportional mode. For processes with a large dead time to time constant ratio (dead time dominant), the integral time for most tuning methods is about ½ the dead time, and the integral mode contribution approaches the proportional mode contribution for minimizing peak error. For lambda tuning of dead time dominant self-regulating processes, the integral time is set equal to the time constant which makes the integral mode contribution much larger than the proportional mode contribution. The addition of the low limit of 0.4 times the dead time for the reset time helps give more proportional action. Note that the controller gain affects the contribution from both modes. For dead time dominant processes the controller gain is less than the inverse of the open loop gain resulting in the peak error approaching the open loop error. Equation C.2 (Appendix C) shows how the ultimate limit to the peak error depends upon the ratio of the dead time to the 63% response time (dead time plus primary time constant).

For a proportional-integral (PI) controller:

$$E_x = \frac{1.5}{K_c * K_o * \left(1.0 + \dfrac{0.5 * \theta_o}{(T_i + \Delta t_x + \tau_f)}\right) + 1.0} * E_o \tag{1.17}$$

For processes with a small time constant to dead time ratio (dead time dominant), the product of the PID gain and open loop steady state gain is about 0.2 and the reset time is less than ½ the dead time. The denominator ends up being about equal to the numerator. The result is that the peak error is about equal to the open loop error which is consistent with the observation that there is little if any reduction in open loop error by feedback control for an unmeasured step load disturbance to a dead time dominant loop. There is more integral action than proportional action and the process has reached close to its final value in one dead time, the minimum time needed for seeing and correcting for an unmeasured disturbance.

For dead time dominant processes, the peak error for a step disturbance approaches the open loop error (error if there was no feedback control).

For processes with a large primary time constant to dead time ratio, a feedback controller is potentially capable of making the peak error a small fraction of the open loop error if the controller is aggressively tuned to make the most of the proportional and derivative mode to provide an immediate response. The numerator in Equation 1.17 reduces from 1.5 to 1.25 from the derivative mode contribution giving Equation 1.18a.

The peak error is best minimized by the more immediate action provided by the proportional and derivative modes possible with processes that have a large time constant to dead time ratio.

For a proportional-integral-derivative (PID) controller:

$$E_x = \frac{1.25}{K_c * K_o * \left(1.0 + \dfrac{0.5 * \theta_o}{(T_i + \Delta t_x + \tau_f)}\right) + 1.0} * E_o \qquad (1.18a)$$

For PID control of processes dominated by a large primary time constant tuned with an aggressive gain as a near-integrator to minimize peak error, the equation simplifies to:

$$E_x = \frac{1.1}{K_c * K_o} * E_o \qquad (1.18b)$$

The size of the peak error for near-integrating, true integrating and runaway processes is inversely proportional to the PID gain.

Nomenclature for Peak Error Metric:

E_x = peak error from load disturbance (e.u.)
E_o = open loop error (load disturbance error with controller in manual) (e.u.)
K_c = controller gain (dimensionless)
K_o = open loop self-regulating process gain (%/%) (dimensionless)
T_i = integral time (reset time) (sec)
τ_f = signal filter time constant (sec)
Δt_x = controller execution time (sec)

1.4.3 Load Disturbance Lag

A disturbance lag (time constant) does not have as much an effect on the integrated error as on the peak error. A disturbance lag that is much greater than the dead time offers a considerably smaller peak error. However, the accumulated error is about the same because of the long protracted return to setpoint. A disturbance dead time does not affect the peak or integrated errors but does affect feedforward timing.

A disturbance lag much greater than the dead time will greatly reduce the open loop error seen within one dead time and consequently significantly decrease the peak error if the PID is tuned aggressively to take most of its action in one dead time.

If we consider a load disturbance that is not a step change but has a time constant, the peak error is reduced to the fraction of the load change in one dead time. We can approximate the effect of a load time constant (τ_L) as the exponential response of the open loop error in one dead time giving Equation 1.19a.

$$E_x = \frac{1.25}{K_c * K_o * \left(1.0 + \frac{0.5 * \theta_o}{(T_i + \Delta t_x + \tau_f)}\right) + 1.0} * (1 - e^{-\theta_o/\tau_L}) * E_o \quad (1.19a)$$

For PID control of processes dominated by a large time constant (the case of more prominent interest), the equation simplifies to:

$$E_x = \frac{1.1}{K_c * K_o} * (1 - e^{-\theta_o/\tau_L}) * E_o \quad (1.19b)$$

The most common process disturbance is a change in process or utility flow. The change in flow setpoint may be a step but the change in PID output is gradual because integral action is used more than proportional action in flow loops. Loops where proportional action dominates have a high controller gain because of the existence of a large process time constant (e.g., near-integrating processes) to provide enough muscle to preemptively correct a process that is slow to respond. Consequently the response of loops on fast and slow processes can be characterized by a closed loop time constant (τ_c) that is set as the lambda parameter in the lambda tuning method for self-regulating processes. The load disturbance time constant can be taken as being the lambda (λ_L) of the loop that is the source of the load disturbance. If a different tuning method is used, the lambda self-regulating process tuning rule for PID gain (Equation 1.4d) can be solved for lambda. This equation can be used for disturbances originating from integrating processes by finding the equivalent self-regulating process open loop gain and time constant by the use of Equation 1.3a for the conversion of dynamics between near-integrating and true integrating processes.

The dynamic response of composition, pH, and temperature disturbances as they pass through an intervening volume are delayed and filtered by a process time constant. Well-mixed volumes are split into plug flow and perfectly mixed volumes. The plug flow volume acts as a pure transportation delay (θ_p) and the perfectly mixed volume provides a process load time constant (τ_L). The total residence time minus the plug flow volume dead time (turnover time from agitation and recirculation) is approximately the perfectly mixed residence time and hence the process load time constant (Equation 1.20b).

Equations 1.19a and 1.19b show the effect of a load disturbance time constant on the peak error for given settings. Since the effect of the primary time constant in the affected process is also a factor, Equation C.1 in Appendix C is multiplied by the load time constant factors for the originating loop and the intervening volume. The result is Equation 1.20a for the peak error that includes the effect of both time constants and the tuning of the loops that are the source and destination of the disturbance.

The lambda (λ_L) for the loop that is the origin of the load disturbance (e.g., flow loop) is a closed loop time constant. The lambda (λ_o) for the loop that is the destination of the disturbance is an arrest time since the primary loop considered (e.g., composition, pH, or temperature) has a near-integrating, true integrating or runaway response.

$$E_x = [1 - e^{-(\theta_o + \lambda_o)/\tau_o}] * [1 - e^{-(\theta_o + \lambda_o)/\tau_L}] * [1 - e^{-(\theta_o + \lambda_o)/\lambda_L}] * E_o \qquad (1.20a)$$

For intervening back mixed liquid volumes between the origin and destination for composition, pH, and temperature control the load time constant is the residence time minus the mixing dead time (½ of turnover time):

$$\tau_L = \frac{V_b}{F_f} * \theta_p \qquad (1.20b)$$

The effect on the peak error of primary and secondary loop tuning and the residence time of intervening volumes can be better estimated from the lambdas used in tuning the loops.

Nomenclature for the Effect of Load Disturbance Lag and Lambda on Peak Error:

E_o = open loop error (error with controller in manual) (e.u.)
E_x = peak error from the load disturbance (e.u.)
K_c = controller gain (dimensionless)
K_o = open loop self-regulating process gain (%/%) (dimensionless)
T_i = integral time (reset time) (sec)
F_f = total feed rate to intervening liquid volume (m^3/sec)
V_b = back mixed intervening liquid volume (m^3)

λ_o = lambda (arrest time) of the primary loop affected by the load disturbance (sec)
λ_L = lambda (closed loop time constant) of the loop (e.g., flow loop) that is the source of the load disturbance (sec)
θ_o = total loop dead time (sec)
θ_p = mixing dead time (e.g., ½ the turnover time) (sec)
τ_f = signal filter time constant (sec)
τ_L = load disturbance time constant of intervening volume (sec)
Δt_x = controller execution time (sec)

1.4.4 Setpoint Response Rise Time, Overshoot, and Undershoot

In the normal operation of continuous processes, the setpoint response is of minimal concern because the setpoint is rarely changed. During startup and product type or grade transitions, the setpoint response can become as important as for batch operations.

The time to reach a new setpoint (rise time) in batch and continuous processes is minimized to reduce the time to reach a new operating condition. For near-integrating, true integrating, and runaway processes, the effect on startup, transition, and batch times is significant. However, the rise time is much less important for bioreactors because the batch cycle times are long (e.g., 10 days) and fixed and the consequences of a temperature or pH overshoot are so severe.

> ***The time to reach a new setpoint (rise time) is important for minimizing the startup and transition time of continuous unit operations and the cycle time of chemical batch operations.***

The maximum deviation after the first crossing of the setpoint (overshoot) must be minimized for many of the same reasons as for minimizing peak error. Overshoot is independent of direction and is the maximum deviation above and below setpoint for an increase and decrease in setpoint, respectively. Bioreactors are particularly sensitive to overshoot in batch temperature or pH. An overshoot can result in cell death. In highly exothermic chemical reactors, an overshoot can trigger a runaway reaction.

The maximum deviation after a second crossing of the setpoint (undershoot) increases the settling time to reach the more optimum setpoint, possibly lowering reaction yield from reduced growth and product formation rates for bioreactors and low reaction rates for chemical reactors. Figure 1.4 shows the rise time, overshoot, and undershoot for step increase in setpoint.

Overshoot and undershoot tend to have more of an effect on process efficiency than rise time.

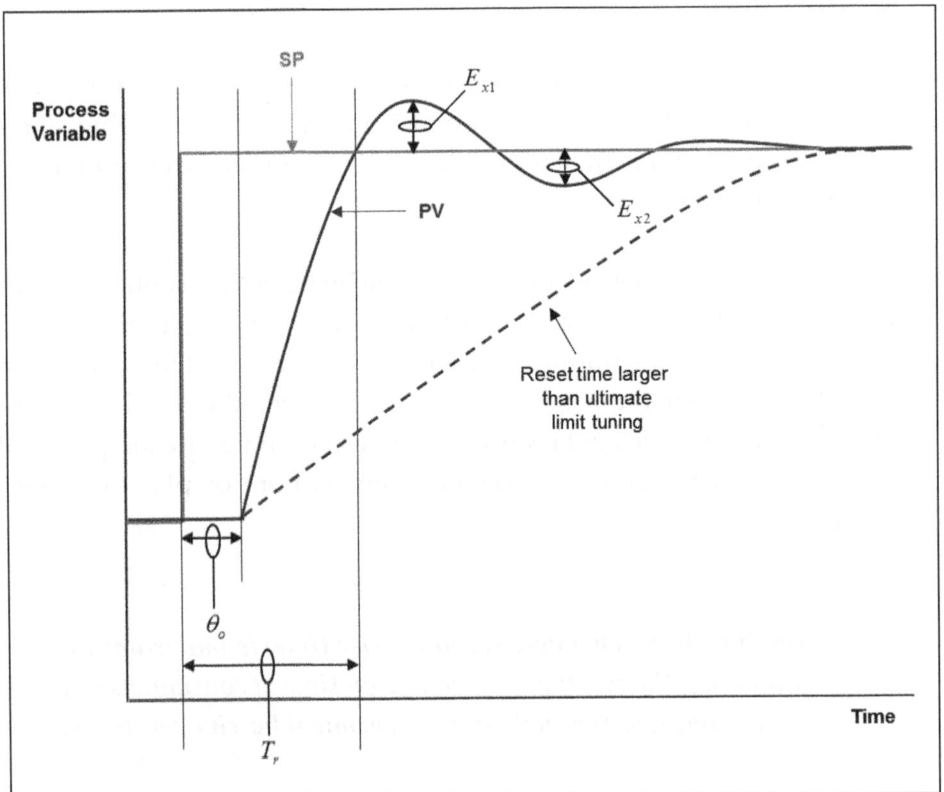

Figure 1.4 – Rise Time (T_r), Overshoot (E_{x1}), and Undershoot (E_{x2}) Metrics for Setpoint Response

The minimum rise time (T_r) can be approximated as the change in percent setpoint ($\Delta\%SP$) divided by the maximum rate of change of the process variable plus the total loop dead time. For an integrating or near integrating process, the maximum percent process variable ramp rate is the open loop integrating process (K_i) gain multiplied by the change in controller output. The change in controller output is the minimum of the maximum allowable change per output limit ($\Delta\%CO_{max}$) and the change as the result of the controller gain (K_c) and the feedforward gain (K_{ff}) as detailed in the denominator of Equation 1.21a. If the step change in controller output from the proportional mode for a structure of proportional action on error is less than the maximum available output change ($\Delta\%CO_{max}$) that is the difference between current output and output limit, Equation 1.21a simplifies to Equation 1.21b for feedback control. In general the tuning and PID structure that minimizes rise time increases the propensity for overshoot and undershoot.

The output change can be proactively set to make the setpoint response faster. For setpoint feedforward, the step change in output is a combination of the feedforward and feedback action. For the smart bang-bang logic detailed by the author in the article "Full Throttle Batch and Startup Response" (*Ref 45*), the step output change is the maximum available output change. These methods offer the opportunity to reduce rise time without increasing overshoot. Setpoint rate limits on the manipulated secondary flow controller with external reset feedback to the temperature controller provide a directional move suppression that can be used to smooth out step changes in flow from utility and raw material systems.

$$T_r = \frac{\Delta\%SP}{K_i * \min(|\Delta\%CO_{max}|, (K_c + K_{ff})*\Delta\%SP)} + \theta_o \tag{1.21a}$$

For a maximum available output change larger than the step from the proportional mode $((|\Delta\%CO_{max}| > K_c * \Delta\%SP))$ the changes in setpoint in the numerator and denominator cancel out yielding a simpler equation:

$$T_r = \frac{1}{(K_i * K_c)} + \theta_o \tag{1.21b}$$

For the near-integrating process response seen in vessel and column temperature loops where the process time constant is significantly larger than the total loop dead time, the integrating gain is the open loop gain (K_o) divided by the open loop time constant (τ_o) yielding Equation 1.20c.

$$T_r = \frac{\tau_o}{(K_o * K_c)} + \theta_o \tag{1.21c}$$

The equations assume that there is a step change in PID output that is the step change in setpoint multiplied by the PID gain. However, these step changes in the PID output only occur if there is no setpoint filter, rate limit, or lead-lag and a PID Structure with full proportional action on error is used (e.g., "PI on error and D on PV" Structure or the beta factor set equal to 1.0 in a 2 Degrees of Freedom Structure). A setpoint lead-lag with the lead time set to zero is the same as a setpoint filter. The promotion of a step change in PID output increases overshoot and undershoot unless the reset time is increased enough that the proportional mode dominates correcting the step change in PID output as the process variable approaches setpoint.

A PID Structure with derivative action on error can help get through large a valve or VSD deadband and resolution limit for small setpoint changes and small gains by a kick in the PID output (step change followed by a rapid exponential decay similar to what is seen when the lead is larger than the lag in a lead-lag block). The result can be an increase in the initial rate of approach to the setpoint. Otherwise, the impact on rise time from derivative action on error is minimal. Normally, a PID Structure with derivative action on PV (e.g., "PI on error and D on PV") or a small gamma in a 2 Degrees of Freedom PID Structure is chosen to minimize kicking in the PID output from the derivative mode for setpoint changes.

PID structures, tuning and features applied to reduce overshoot tend to increase rise time.

Setpoint rise time and overshoot can be minimized for near-integrating, integrating and runaway processes by smart bang-bang logic. The disruption to the loops is minimized by the use of a setpoint rate limit for the manipulated flow and external reset feedback.

Nomenclature for Rise Time Metrics:

$\Delta\%CO_{max}$ = maximum change in controller output converted to % of PID output scale (%)

$\Delta\%SP$ = step change in setpoint converted to % of PID input scale (%)

K_c = controller gain (dimensionless)

K_{ff} = feedforward gain (%/%) (dimensionless)

K_i = open loop integrating process gain (%/sec/%) (1/sec)
K_o = open loop self-regulating process gain (%/%) (dimensionless)
T_r = rise time (time to reach a new setpoint) (sec)
τ_o = open loop time constant (primary time constant) (sec)
θ_o = total loop dead time (sec)

Step changes and oscillations in the PID output can be disruptive for heat integration and to other loops that are relying upon a constant pressure in the same volume (e.g., a utility header or inline system) that is the source or destination of the flow manipulated by the PID. Some tuning techniques are designed to make the changes in the PID output as smooth as possible. Some go so far as to choose to minimize overshoot of the PID output beyond its final resting value (FRV). While this objective may work for volumes with no back mixing and short residence times (e.g., catalyst bed gas reactors and plug flow liquid reactors), preventing step changes in the PID output and overshoot of the FRV is not feasible for integrating processes (e.g., level or batch temperature) and downright dangerous for runaway processes because an excursion in the process variable will not reverse direction until the PID output correction goes beyond the FRV.

While prevention of PID output overshoot is possible for continuous processes, it is not advisable for well-mixed volumes with large residence times that consequently have a large primary process time constant. The rise time can become orders of magnitude larger for these near-integrating processes. A more prevalent goal is to focus on the minimization of process variable overshoot (peak error after first crossing of the setpoint) and undershoot (peak error after second crossing of the setpoint). Undershoot is in the opposite direction of overshoot.

PID output overshoot of the final resting value of the manipulated variable (e.g., jacket temperature or coolant flow) is necessary for an integrating process or runaway process to reach a new setpoint. PID output overshoot is of practical necessity for near-integrating processes to reach a new setpoint in a matter of minutes rather than hours. The objective of minimizing PID output overshoot is only reasonable for moderate self-regulating or dead time dominant processes (e.g., plug flow unit operations).

In general, a decrease in rise time causes an increase in overshoot for feedback control. Overshoot and undershoot of the process variable can be reduced by making the contribution from the proportional and derivative modes optimally larger than the contribution from the integral mode. The integral mode is the principal cause of overshoot because this mode does not reverse direction of the change in PID output until the process variable crosses setpoint. The integral mode has no sense of direction as seen in the Appendix B temperature control example.

Overshoot and undershoot can be eliminated by a PID Structure with no proportional action on error or a setpoint filter time equal to the reset time at the expense of a considerable increase in rise time. The tradeoff between rise time and overshoot can be optimized by the use of a 2 Degrees of Freedom (2DOF) structure and adjustment of the beta and gamma factors or the use of a setpoint lead-lag and an adjustment of the lead time to lag time ratio given that the lag time is set equal to the reset time.

The tradeoff between rise time and overshoot from feedback control can be optimized by the use of 2 Degrees of Freedom PID structure or a setpoint lead-lag.

For batch temperature control where there is only heating or only cooling (no split range), and there is no significant heat generation from reaction or heat loss from evaporation in the process, the response is in one direction only. For heating the temperature can only increase and for cooling the temperature can only decrease. A one-sided response (unidirectional response) also occurs in batch pH control where there is only an acid or only a base reagent and there is no significant reagent consumption from reaction. For these one-sided responses, integral action will cause overshoot. The structure must have proportional action on error, derivative action on the process variable and no integral action. When integral action is not used, an output bias must be specified. For these one-sided batch applications, the output bias is generally the PID output when the process variable is at setpoint.

For batch control loops where the process variable can only respond in one direction, a PID structure with P on error and D on PV with no I must be used to prevent overshoot. The PID output bias is the PID output when the PV is at setpoint.

2 Inferential Measurements

2
Inferential Measurements

Inferential measurements can provide fast online prediction of key variables that are indicative of reactor performance. The variables are usually compositions or performance indicators based on compositions.

The inputs to the inferential measurements use sensors, probes, and meters that can be inserted into a line or a vessel. The measurements are continuous and require no sample preparation. The computational time is less than a second except for spectral calculations. As a result, inferential measurements are essentially as fast as the sensors.

> ***Inferential measurements can be essentially as fast as the sensors. The performance of process control systems using inferential measurements can be greatly improved by the elimination of dead time from sample transportation, cycle time, analysis time, and multiplex time.***

The predictions should always be corrected by composition measurements from at-line or offline analyzers and first principle calculations. The correction is best done automatically by simply adding a fraction (e.g., 0.4) of the error that is the analysis or computation result minus the synchronized inferential measurement. The inferential measurement is passed through dynamic blocks such as dead time and filter blocks for synchronization (dynamic compensation) in continuous processes. Section 2.5 on Step Response Models discusses this correction. For batch end points, no synchronization is necessary and the correction is simply done after the batch is completed and the analysis result is communicated.

A definitive prediction of a future value can also be obtained from the change in an inferential measurement with time. The prediction of a future value can be realized by computing the slope of the inferential measurement profile and then multiplying the slope by an appropriate time interval into the future. For

Inferential measurements of batch endpoints do not need the dynamic compensation required by inferential measurements of stream compositions in continuous processes for synchronization and correction with analyzer results.

batch operations, the slope of the batch profile is used. For continuous operations, the slope of the approach to a new setpoint is used. This future value is useful for predicting when a key variable will reach a new operating point during the startup and transition of continuous processes and a new setpoint or the endpoint of batch processes. For sensitive measurements of excess reactants, just a large change in sensor output can be used to indicate a batch has reached its endpoint (e.g., a large change in conductivity for a batch chlorinator). Section 2.3 on batch endpoints will go into the details of this simple calculation including the crucial use of a dead time block to provide immediate updates with a good signal to noise ratio.

Future values of inferential measurements can be readily computed from the rate of change of a measurement with respect to time (the slope of the trend chart plot) to predict reaching an endpoint for batch operations and a new operating point for all operations. The use of a dead time block to create the slope is crucial as described in the batch endpoint section.

2-1. Conversion

A computation of the heat removal rate in a cooling system with a sensible heat correction can provide an inferential measurement of conversion rate for exothermic reactors. If there are no significant side reactions, the conversion rate can be taken as being specific to the production rate. Figure 2.1 shows the jacket temperature and flow measurements and computations needed. If the jacket circulation rate is constant and known, a flow measurement is not necessary. The description here is for a jacket but the same principles and considerations apply to a coil.

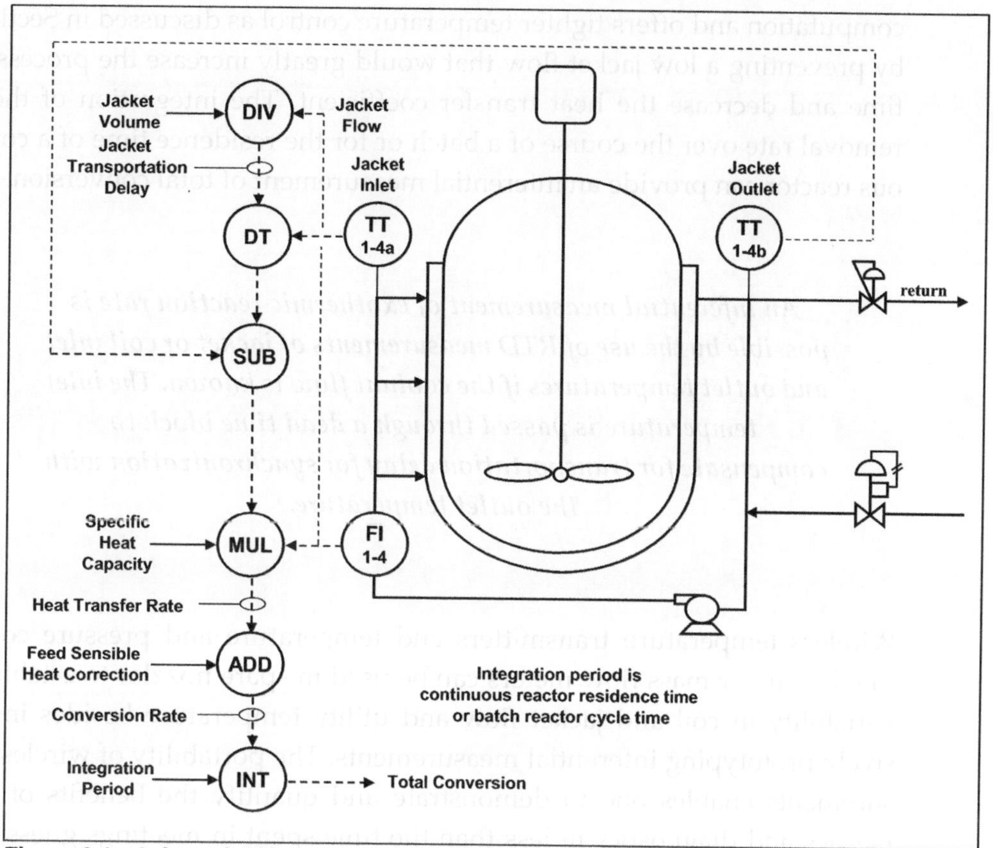

Figure 2.1 – Inferential Measurements of Exothermic Conversion Rate and Total

The heat removal rate uses temperature sensors on the inlet and outlet of the jacket. The inlet temperature measurement is sent through a dead time block with a dead time setting equal to the transportation delay so the inlet temperature can be synchronized with the outlet temperature.

Resistance temperature detectors (RTDs) provide a more accurate inferential measurement than thermocouples that is critical to the computation of small differences between the inlet and outlet temperatures. The heat removal rate is the outlet temperature minus the synchronized inlet temperature multiplied by the jacket flow and the heat capacity of water at the operating temperature. If the jacket flow is not constant, the heat removal computation and the transportation delay must be updated based on a jacket flow measurement. For a constant coil or jacket flow and cooling or chilled water temperature, the difference between the reactor and coil or jacket outlet temperature (approach temperature), can be used as a simpler computation. The use of a jacket recirculation system to maintain a constant jacket flow simplifies the computation and offers tighter temperature control as discussed in Section 6.1 by preventing a low jacket flow that would greatly increase the process dead time and decrease the heat transfer coefficient. The integration of the heat removal rate over the course of a batch or for the residence time of a continuous reactor can provide an inferential measurement of total conversion.

An inferential measurement of exothermic reaction rate is possible by the use of RTD measurements of jacket or coil inlet and outlet temperatures if the coolant flow is known. The inlet temperature is passed through a dead time block to compensate for transportation delay for synchronization with the outlet temperature.

Wireless temperature transmitters and temperature and pressure compensated annubar mass flow meters can be used in spare nozzles to establish the variability in coil and jacket flow and utility temperature besides inexpensively prototyping inferential measurements. The portability of wireless measurements enables one to demonstrate and quantify the benefits of online metrics and diagnostics in less than the time spent in meetings guessing the value and feasibility.

Wireless devices offer portability and ease of installation for rapid, low cost demonstration of the value of inferential measurements.

The sensible heat correction can be readily computed. The net sensible heat is computed for each feed stream as the reactor temperature minus the stream temperature multiplied by the stream flow rate and specific heat. The feed sensible heat correction is the summation of the net sensible heats for all feed streams. The feed streams are typically cooler than the temperature of an exothermic reactor. Part of the heat of reaction goes into bringing the feed added up to the operating temperature. The cooling effect of these streams is added to the heat removal rate to arrive at a better estimate of the exothermic reaction rate. If the feed streams are hotter than the reactor, the sign of the correction is negative.

2-2. Composition

The inferential measurement of the concentration of a component (e.g., a reactant or reagent) in a feed, exit or recycle stream or vessel is possible if the component of interest has a sufficiently large, definitive effect on conductivity, dielectric spectroscopy, density, near infrared (NIR) spectroscopy, pH, turbidity, or viscosity. To infer composition in vessels, sensors are installed in nozzles in the vessel, in recirculation lines, or in sample lines in close proximity to the vessel.

Online or inline measurements such as conductivity, dielectric spectroscopy, density, near infrared (NIR) spectroscopy, pH, turbidity, and viscosity have been used to provide inferential measurements of current and future values of stream or vessel compositions.

If the component of interest has a higher conductivity than other components or the solvent, conductivity can be used if the conductivity is restricted to be always on one side of the peak in the conductivity curve. The process gain changes sign as an operating point crosses the peak, which is disastrous to a control loop because the PID control action sign must be the opposite of the process action sign (assuming that valve action is correctly configured). If the PID control action has the wrong sign, the PID output will ramp to its output limit.

Plots of conductivity versus concentration have a peak if the entire concentration range is covered. Note that conductivity versus concentration curves are a function of temperature. Also at the peak there is a dramatic decrease in the slope and consequently a loss of sensitivity of the conductivity to changes in concentration. The process concentration must not get within a few percent of the peak.

A few ionic species such as sulfuric acid have two peaks. The second peak for sulfuric acid occurs at 93% concentration. Since most plants use 98% sulfuric acid for reagent, reagent concentration control by the use of conductivity is only feasible if a concentration of less than 95% is physically impossible.

For product conductivity that traverses the peak, concentration control cannot start until the conductivity is well to the right of the peak. Figure 2.2 shows a plot of conductivity for a concentration range of 0–60% for common salts and strong acids and bases. For a very low concentration of a single acid or base (e.g., < 0.01% by weight) pH provides a more sensitive inferential measurement of concentration.

Conductivity is used to infer the concentration of a single acid, base, salt, or non-aqueous solvent. The concentration must not cross a peak in the conductivity curve.

pH is used to infer a concentration less than 0.01% by weight of a single acid or base.

For conductivity and pH, a signal characterizer is commonly used to compute the X axis (e.g., salt, acid, or base ion concentration) from the Y axis (conductivity or pH). The 20 or more data points of a signal characterizer block are usually enough to provide a piecewise linear fit. The points are more closely packed in the operating regions of greatest interest or nonlinearity. A cascade of characterizers can be used to provide greater resolution where a secondary signal characterizer is added to the output of the primary characterizer.

Since these curves depend upon temperature, the samples used to generate the curve should be at the normal operating temperature. A simple temperature correction should be generated based on the results of samples that cover the entire possible range of temperatures including abnormal operation. The correction is typically done by subtracting a reference temperature from the operating temperature and multiplying the result by a factor. The corrected pH or conductivity then becomes the input to the signal characterizer. Note that samples taken must use the same equation for temperature correction. Since lab meters often do not have this capability (e.g., solution pH temperature correction), which is commonly available in smart transmitters, a wireless smart transmitter is recommended for lab measurements.

For a two component liquid mixture where the density of the components differs by more than 1%, an extremely accurate concentration measurement is attainable by the use of a Coriolis meter. The accuracy of the density measure-

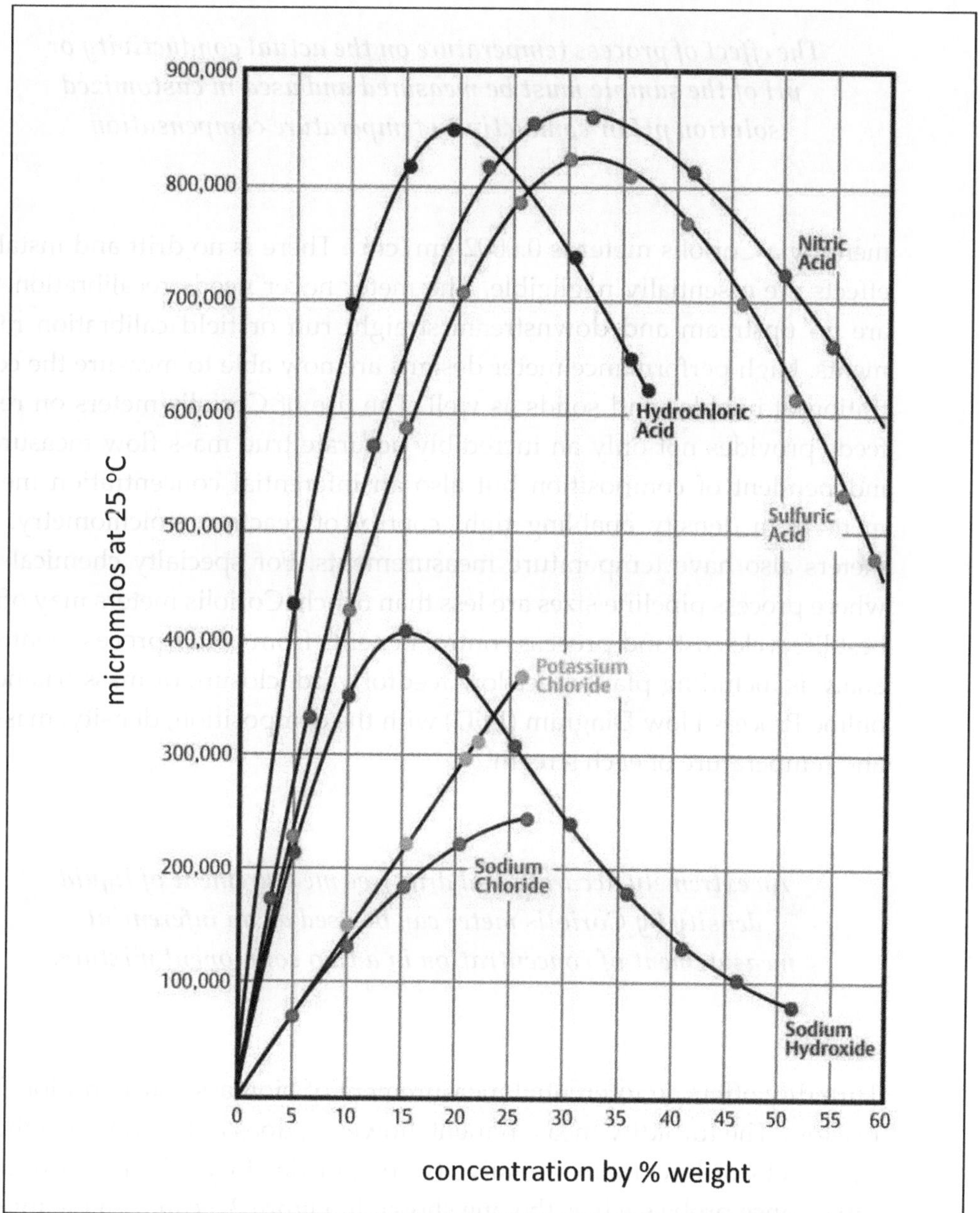

Figure 2.2 – Inferential Measurement of Concentration from Conductivity

One or more signal characterizer blocks are used to give a piecewise linear fit that enables the online computation of the X axis (% concentration) from the Y axis of a conductivity curve or pH titration curve generated from samples at various operating conditions.

> ***The effect of process temperature on the actual conductivity or pH of the sample must be measured and used in customized solution pH or conductivity temperature compensation.***

ment by a Coriolis meter is 0.0002 gm/cm^3. There is no drift and installation effects are essentially negligible. The meter never needs recalibration. There are no upstream and downstream straight run or field calibration requirements. High performance meter designs are now able to measure the concentration of bubbles and solids as well. The use of Coriolis meters on reactant feeds provides not only an incredibly accurate true mass flow measurement independent of composition but also an inferential concentration measurement from density enabling tight control of reaction stoichiometry. These meters also have temperature measurements. For specialty chemical plants where process pipeline sizes are less than 6 inch, Coriolis meters may offer the best lifecycle cost and process control benefits from better process control and analysis including plantwide flow feedforward, closure of mass balances, an online Process Flow Diagram (PFD) with the composition, density, mass flow, and temperature of each stream.

> ***An extremely accurate and drift free measurement of liquid density by Coriolis meter can be used as an inferential measurement of concentration in a two component mixture.***

Turbidity offers an inferential measurement of biomass concentration for bioreactors. The turbidity measurement, however, does not distinguish the number of cells or how many are alive (viable) or dead (lysed). The addition of a capacitance probe such as the one shown in Figure 2.3 can provide inferential measurements of mammalian cell size, homogeneity, and membrane integrity. Viable mammalian cells have membranes intact whereas lysed cells have holes or fractures in their membranes.

The computation of an oxygen uptake rate (OUR) can provide an inferential measurement of cell growth rate. Because the Dissolved Oxygen (DO) PID output is continuously adjusted to match the amount of oxygen taken up by cells, the OUR can be roughly estimated from the air or oxygen flow demand

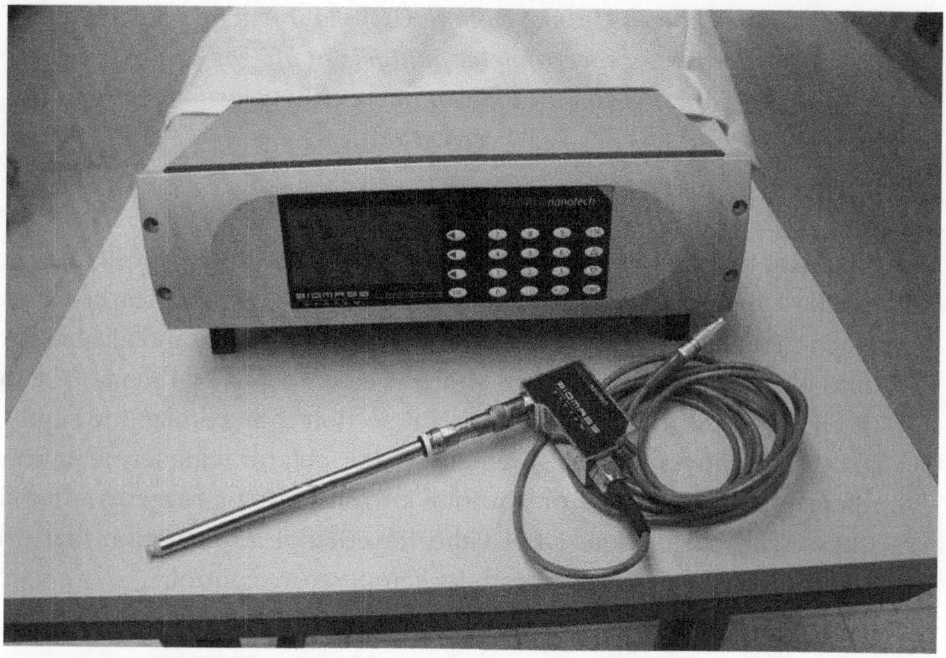

Figure 2.3 – Inferential Measurements of Cell Size, Homogeneity, and Membrane Integrity from Dielectric Spectroscopy (Fogale Nanotech)

Dielectric spectroscopy in combination with an inferred measurement of biomass from turbidity can provide inferential measurements of cell size, and the relative concentration of viable (live) versus lysed (dead) cells.

when a dissolved oxygen (DO) controller is in automatic. Alternately, the air flow can be shutoff and the rate of dissolved oxygen decay multiplied by the bioreactor volume provides an indication of OUR.

A more sophisticated OUR calculation uses a mass spectrometer to measure the amount of oxygen in the off-gas. The oxygen flow rate departing the bioreactor is the off-gas oxygen concentration multiplied by the vent gas flow. The oxygen flow rate entering is the sparge gas oxygen concentration multiplied by the sparge flow. Note that some bioreactors use oxygen besides air in the sparge to meet peak oxygen demand; others use inert gas in the sparge to sweep dissolved carbon dioxide out of the bioreactor.

The Oxygen Uptake Rate (OUR) in bioreactors computed from mass spectrometer analysis of off-gas and oxygen addition rate can provide an inferential measurement of cell growth rate.

The negative rate of change of weight measured by load cells has been used by the University of California – Sacramento to compute carbon dioxide production rate (CPR) as an indicator of yeast growth rate and alcohol production rate for wine. The weight loss should be computed by the use of a dead time block as noted in the batch endpoint section to maximize the signal to noise ratio. This inferential measurement works well for wine fermentation because there is no sparge gas or addition of nutrients or reagents after the batch starts. This calculation is not valid for ethanol fermentation that uses direct injection of steam into the broth for temperature control.

The Carbon Dioxide Production Rate (CPR) in alcohol fermenters with no sparge gas or nutrient or reagent addition can be computed from load cell computation of weight loss. CPR can provide an inferential measurement of yeast growth rate and alcohol production rate.

CPR and OUR can be used for diagnostics and prediction of batch endpoint. The use of data analytics software to detect and diagnose abnormal batches through principal component analysis (PCA) and to predict product quantity and quality through projection to latent structures (PLS) benefits from inputs such as CPR and OUR that have embedded process knowledge. OUR can also be used as a feedforward signal for glucose and glutamine control loops.

OUR and CPR provide diagnostics and predictions as inputs to data analytics software.

A virtual plant using first principle chemical engineering models running real time can provide inferential measurements of key component concentrations.

Model predictive control (MPC) or valve position control (VPC) can adapt the virtual plant model parameters for utilization rates by matching virtual to actual PID outputs. At-line or offline lab analyzers can be used to adapt parameters for formation rates by matching actual to predicted slopes of concentration profiles. Knowledge gained through model and research data can also lead to manual adaptation of model parameters. A VPC that uses an enhanced PID developed for wireless (see Appendix E) can deal with discontinuities, noise, and interactions during adaptation of the virtual plant.

> *A virtual plant running real time adapted online by MPC or VPC or manually to match the actual plant can provide inferential measurements of key component concentrations.*

NIR spectrometers have been used successfully to predict the ethanol yield of corn. In ethanol manufacturing, an inferential measurement of fermenter yield can provide an immediate reduction in the largest cost and source of variability by automatically adjusting the corn feed rate. The corn yield factor multiplied by the mass feed rate of corn from a gravimetric feeder is used to create an inferential measurement of production rate measurement for PID control. Section 5.4 on bioreactor (ethanol fermenter) control and optimization shows how the simple configuration of a PID for front end production rate control not only achieves an immediate optimization but also provides a handle for operations personnel to coordinate ethanol plant front end and back end production rates.

> *NIR spectrometers of small samples from the corn feeder can predict the ethanol yield to reduce the largest cost and source of variability in ethanol production.*

2-3. Batch End Point

The slope of a batch profile can be estimated from periodic at-line or scheduled off-line analyzer results by inputting each new analysis into a dead time block. The output of the dead time block that is the old analysis is subtracted from the new analysis to get a change in concentration that when divided by the dead time is the slope of the concentration profile. The time interval chosen as the dead time parameter must be large enough to ensure that the difference between the new and old values is much larger than noise. The dead time block provides a smooth and fast continuous update to the slope with a good signal to noise ratio. Techniques that compute a rate of change on a discontinuous basis with an update period equal to the time interval over which data is gathered create a dead time equal to ½ the period and introduce abrupt changes.

> ***The simple use of a dead time block to compute the rate of change of a process variable provides a smooth and fast update to the batch profile slope with a good signal to noise ratio.***

The slope can be used for profile control as described in Section 5.8 on biological fed-batch profile control or used to predict the batch endpoint. From the slope near the end of the batch, the additional product produced in the dead time bock time interval or the analysis time interval for sampled measurements is computed. The slope (e.g., the conversion rate, cell growth rate, product formation rate) near the end of the batch is used to make an economic decision as to whether the batch cycle or phase should be terminated for extra capacity or extended for extra yield.

Here we use a more generic definition of a batch endpoint as being a point at the end of a batch phase and not necessarily the end of the total batch. The slope is converted to product mass flow and multiplied by the analysis time interval to get the additional product mass for the given dead time or analyzer time interval. The current product mass in the batch divided by the mass of each key raw material added in the batch gives the yield in terms of product for each key raw material. The additional product mass per batch is divided by this yield and multiplied by the cost per unit mass of each key raw material. The results are summed to arrive at a dollar value of the increased yield by extending the batch phase or cycle time.

The current product mass in the batch divided by the current batch time in hours offers an estimate of the current batch production rate. Alternately the production rate can be determined from the flow controller based on predicted feed yield. The production rate multiplied by the profit per unit mass and finally multiplied by the dead time or analysis time interval gives an estimate of the value of additional capacity made available by terminating the batch.

The inferential measurement of profile slope used in a prediction of endpoint can determine the economic tradeoff between yield and capacity from terminating the batch earlier or later.

The analysis time interval should be shortened to be just large enough for a good signal to noise ratio near the end of the batch to make the optimization more accurate. If the analysis of the profile of a key composition or product is not available until after the batch has been transferred, the results can be cautiously used for the next batch.

A worm plot as shown in Figure 2.4 can be used to predict when a batch is going to be outside the quality specifications. The worm plot consists of batch endpoints plotted as scores versus principal components from multivariate statistical models. Outliers are removed. The points are distinguished by numbers or data point size to show the progression of batches as a worm. Is the worm coiled as desired within the batch specifications shown as a circle or ellipse, or are the latest batches uncoiling and is the head approaching the specification limits?

A worm plot provides pattern recognition of where batches have been and are now headed.

2-4. Multivariate Statistical Models

Multivariate statistical models for continuous processes currently rely on the insertion of dead time in process inputs to model the effect of dynamics between the process inputs and downstream process outputs for synchroniza-

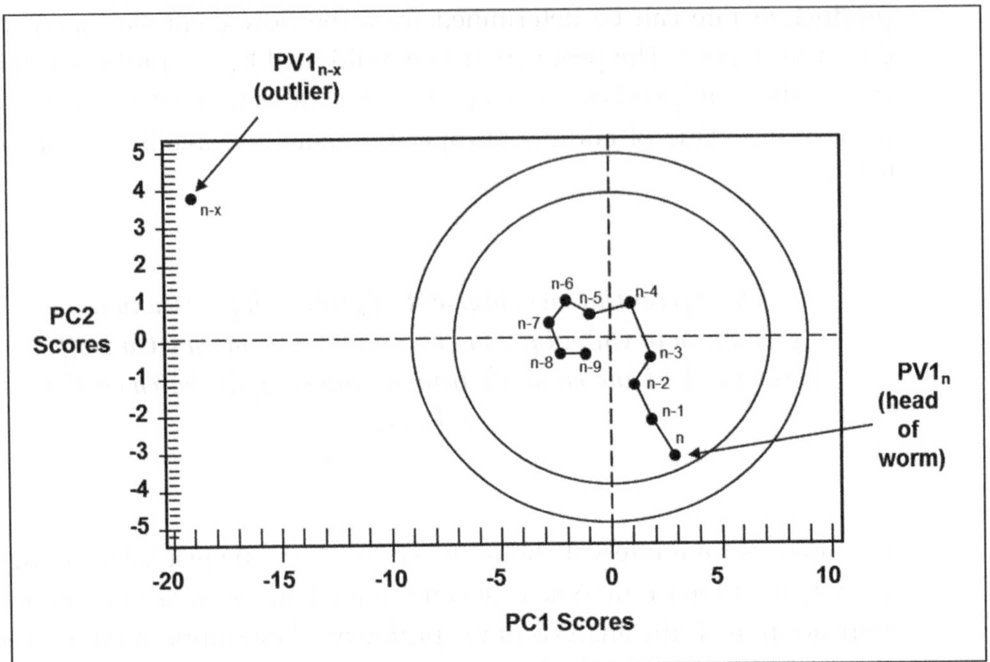

Figure 2.4 – Worm Plot of Batch End Points Shows Where Batch Is Headed

tion with lab results. The dead time is inserted so that the effect of the upstream input is matched to the downstream output or analyzer result. A simple synchronization inserts a dead time that is the residence time (volume/flow) between a downstream process output and an upstream process input plus analyzer dead time. For plug flow volumes (no back or axial mixing) the synchronization is more accurate because the residence time is a simply a pure transportation delay. For well-mixed volumes, the residence time is the sum of a mixing delay and a process time constant corresponding to the 63% response time of the change in the downstream output to a change in the upstream input.

A dead time is inserted in each process input that is the residence time in the volume between the input and downstream process output plus analyzer dead time. This synchronization is more accurate for plug flow volumes than well-mixed volumes.

For an at-line analyzer, the dead time is the sample transportation delay plus 1½ times the analyzer cycle time plus ½ the multiplex time. For offline analyzers, the dead time is ½ the time between samples plus transport time to the lab plus the time after a sample arrives in the lab to when a lab result is entered into the software system. Unless procedures are enforced for sample frequency and transport by operators and for sample analysis and data entry by lab technicians, the synchronization of inferential measurement with lab results is inadequate for continuous processes. At best, the updating of inferential measurements from off-line analyzers is slow.

The dead time for an at-line analyzer is the sample transportation delay plus 1½ times the analyzer cycle time plus ½ the multiplex time.

The dead time for an offline analyzer is ½ the time between samples plus the transport time to the lab plus the time for lab analysis and data entry.

The dead time from offline analyzers can be so unpredictable and large that the prediction for continuous processes is unusable due to the lack of timely and accurate correction by analysis.

For batch processes, the prediction of batch endpoint conditions does not require the insertion of dead times on the inputs. The effect of each input is correlated through principal components to a projection to latent structures (PLS) output at one point in time (e.g., batch end point). The nonlinearities of a batch profile are handled by a piecewise linear fit. Tests on data analytics have confirmed this supposition.

While batch processes normally pose more challenges than continuous process in modeling and control due to the lack of a steady state, the persistent process nonlinearities and synchronization problems in continuous processes make the use of data analytics for predictions more difficult.

The data analytics software used to create these models offers the ability to use a large number of process inputs and to reduce the set of inputs to a small

The use of PLS to make predictions is easier for batch than for continuous operations; batch data analytics do not require the insertion of dead time blocks, and the piecewise linear fit of the batch profile handles process nonlinearities.

number of principal components that are independent (inputs are not correlated). In the process of doing the principal component analysis (PCA), the contribution of individual process inputs can be viewed by a drill down. Even if the PCA is not taken to the next step of a prediction by a projection to latent structure (PLS) model, the relative importance of particular process measurements can be analyzed. There are often surprises in that an overlooked process condition may turn out to be the principal cause of a deterioration in product concentration.

PCA and PLS models are useful for analyzing and understanding process performance by discovering unsuspected relationships even if not used for online inferential measurements.

The piecewise linear fit of a nonlinear batch profile is often sufficient. Also, the use of linear projection provides values that are just an extension of the last predictions when the input values are outside of the original data used to develop the models. Even so, as with any data driven model, a correlation between an input and an output may not be a cause and effect relationship. While PCA helps eliminate correlations between inputs used in the PLS model, there may be hidden coincident relationships. A knowledgeable process engineer working with operators must select the batches for model development and must review the PCA and PLS results. Unrepresentative batches and PLS results must be removed as outliers.

Knowledgeable process engineers and operators must eliminate batches used in model development and predicted outputs that are not representative of process operation.

Neural networks offer nonlinear models but the lack of the ability to automatically eliminate correlated inputs and to drill down to analyze individual contributions reduces their utility. The potential of extremely nonlinear predictions requires extra care to prevent buzzing (cycling of the PID back and forth across a dip in the process response curve) or windup of the PID output, both of which can occur when inputs outside of the test data range and reversals of slope sign occur causing a reversal of the sign of process action. Neural networks used to develop continuous process models require the same synchronization with downstream process outputs and lab results described for PLS models.

Neural network models are more susceptible to process gain reversals and unreasonable nonlinear projections when inputs are outside the test data range.

The use of neural network models can be greatly extended by the use of principle components as neural network inputs, a complete design of experiments, the use of corresponding limits in model inputs, dynamic compensation that uses a time constant plus a dead time to match the 63% response time, and frequent analyzer feedback correction.

2-5. Step Response Models

Step response models can take advantage of the software used to identify the dynamics for model predictive control (MPC). The process inputs that affect the inferential measurement are stepped in both directions and if possible with a pseudo random binary sequence similar to what is done in identifying the MPC dynamics of disturbance variables. Analyzer measurements with the same cycle time and analysis time as during normal operation are the process outputs being predicted.

So that the MPC inputs are truly independent variables, controllers that manipulate process inputs to the model and whose controlled variable is affected by the inferential measurement must be in manual. For example, a distillation column temperature controller that manipulates a reflux to feed ratio would need to be in manual during the identification of the dynamics of an inferential measurement of overhead or bottoms composition. The dynamics identified are just used to synchronize inferential measurements with analyzer predictions. The inferential measurements used for process control provide a new update without the delay or lag by omission of the dynamics identified. MPC matrix condition number and analysis can be used to help eliminate correlations between inputs to ensure the MPC inputs are independent variables.

Controllers that would change MPC model inputs as a result of a response to MPC model outputs must be in manual. This practice, plus matrix condition number and analysis, can help ensure that MPC inputs are truly independent variables.

Step response models use an open loop gain, a total loop dead time, and a primary and possibly a secondary time constant. The step response models described here all work in percent signals since PID algorithms work in percent signals. The conversion from percent signals to the engineering units seen in the control room is done by scaler blocks that use the input and output scale ranges.

An open loop steady-state process gain is used for processes that decelerate to a steady state as a result of negative feedback within the process. An open loop integrating process gain is used for processes that ramp as a result of no feedback within the process. An open loop runaway process gain is used for processes that accelerate as a result of positive feedback within the process. The input is a change in the correction variable and the output is the change in the process variable. The models identified by tuner and MPC identification software take into account the controller scale ranges of the manipulated and process variables and include the effects of valve or variable speed drive (VSD) and measurement dynamics. As a result, the process gain identified is really an open loop gain that is the product of the valve or VSD gain, process gain, and measurement gain. Correspondingly, the process dead time is actually a total loop dead time including the dynamics of the automation system. While the primary (largest) and secondary (second largest) time constants are normally in the process for composition and temperature control, they can be in the automation system for flow, level, and pressure control.

A linear dynamic estimator (LDE) for a self-regulating process uses an open loop steady-state process gain, and subsequently a dead time, primary process time constant, and secondary time constant for synchronization with the MPC output.

Figure 2.5a shows the time domain block diagram for a step response model for a self-regulating process. The normal operating value of the percent analysis variable ($\%AV_o$) is subtracted from the new value of the percent analysis variable ($\%AV_n$) to create a percent analysis deviation variable ($\Delta\%AV$) that is added to the change in the disturbance value ($\Delta\%DV$) to get the percent change in the correction variable as a process input ($\Delta\%XV$). The step response output is the normal operating value of the percent process variable ($\%PV_o$) added to a deviation variable ($\Delta\%PV$) that is the output of the series of dynamics blocks and is passed through a scaler block to provide a new value of the process variable (PV_n) in engineering units. The dynamics inserted after the open loop steady-state process gain are a dead time block for the total loop dead time (θ_o) and a filter block for the secondary time constant (τ_s) to provide all of the parameters for a second order plus dead time step response model for self-regulating processes. If the analysis variable equals its normal operating point plus any compensation for the change in disturbance variable, the process variable will settle out to equal its normal operating point. The input and output biases enable the setting of normal operating points. For more than one input, the results of the model for each input are summed to create the new and future values of the process variable.

The future value for a self-regulating process is the correction value multiplied by the open loop steady-state process gain to give a change in the percent future value ($\Delta\%PV_f$). This change is added to the normal operating point and passed through a scaler block to provide a future value of the process variable (PV_f) in engineering units.

Figures 2.5b and 2.5c show the step response model for integrating and runaway processes, respectively. For integrating and runaway processes, an integrator block is substituted for the filter block for the primary time constant. For integrating process models, the normal operating point of the analysis variable (e.g., manipulated flow) that is the negative bias must be sufficiently greater than zero to provide negative as well as positive changes in the process variable. This bias represents a process load. To achieve a new target or

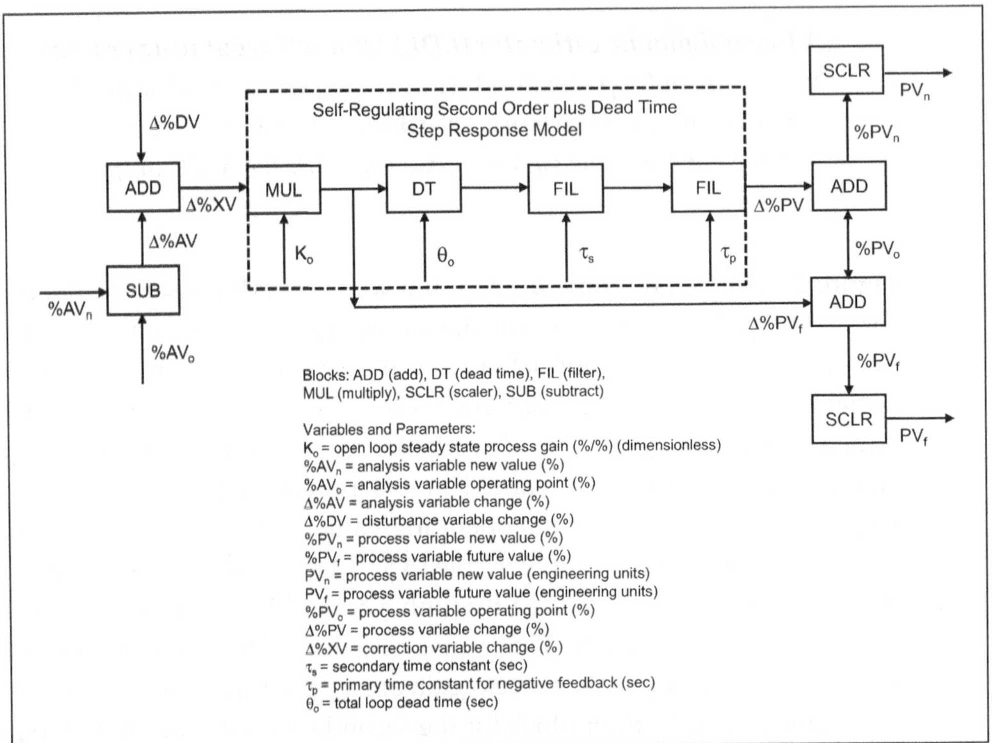

Figure 2.5a – Step Response Model for Self-Regulating Process

> ***The future value of the inferential measurement to be seen in an analyzer result is the instantaneous LDE result corrected by the addition of a fraction of the error between the synchronized LDE output and the measured process output.***

setpoint the analysis variable must be temporarily changed to be different from the load. When the process variable settles out at the new operating point, the analysis variable returns to be equal to the load plus any compensation for the change in the disturbance variable.

> ***A linear dynamic estimator (LDE) for an integrating process uses an open loop integrating process gain, and subsequently a dead time and secondary time constant for synchronization with the MPC output.***

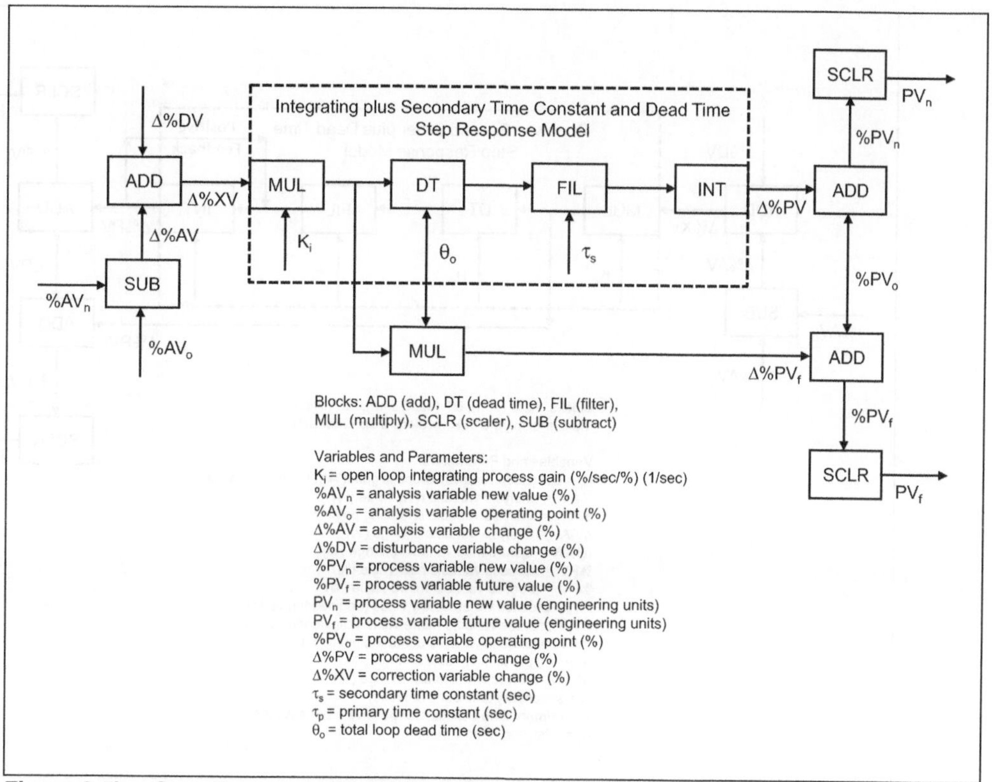

Figure 2.5b – Step Response Model for Integrating Process

> ***A linear dynamic estimator (LDE) for a runaway process uses an open loop runaway process gain, and subsequently a dead time and secondary time constant for synchronization with the MPC output. Runaway processes are often modeled as integrating processes.***

The future value for an integrating process is the correction value multiplied by the open loop integrating process gain and the total loop dead time to give a change in the percent future value ($\Delta\%PV_f$). This change is added to the normal operating point and passed through a scaler block to provide a future value of the process variable (PV_f) in engineering units.

Figure 2.6 shows a linear dynamic estimator that uses the step response model output and future value. The new value of the process variable in engineering units that includes the dynamics of the step response model is compared to the actual measurement (e.g., an at-line analyzer result). The new value for the

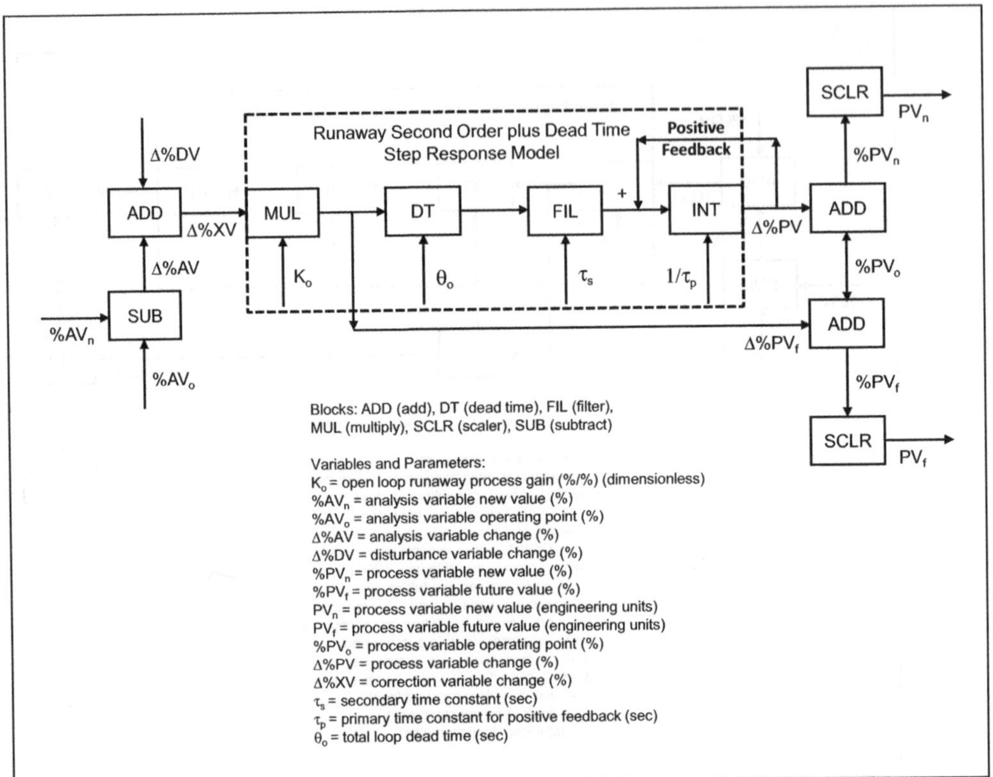

Figure 2.5c – Step Response Model for Runaway Process

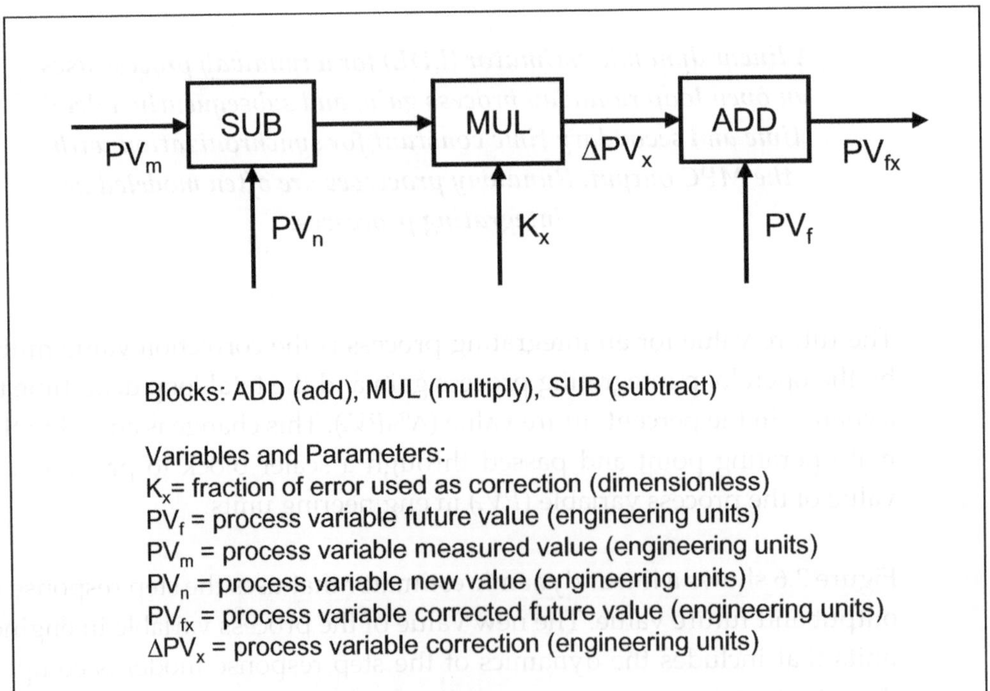

Figure 2.6 – Linear Dynamic Estimator (LDE)

> ***The future value for an integrating process is the LDE instantaneous ramp rate multiplied by a time interval that is dead time plus a secondary time constant used for synchronization.***

step response model output subtracted from the actual measurement is the error in the prediction. The error is multiplied by a factor less than 1.0 (e.g., $K_x = 0.4$). The resulting fraction of the error is added to the future value in engineering units (PV_f) to provide a corrected future value (PV_{fx}).

2-6. First Principle Models

For steady-state first principle models the derivative of the ordinary differential equations (ODE) is set equal to zero and a search of process outputs is conducted to satisfy the resulting ODE. Volumes are treated as perfectly mixed (no turnover time and completely uniform). Profiles are created by partitioning the volumes into subsections. Partial differential equations provide a more detailed analysis than ODE of changes in process temperature and composition depending upon location within the volume and are principally used to study mixing patterns. First principle models iterate until a final solution is reached, giving the final condition of process output streams that satisfy charge, material, momentum, and energy balances and equilibrium relationships for a given set of static process input streams.

These first principle models have all of the equipment and process parameters needed for rigorous design or optimization of operating points. The transition between operating points is as fast as the convergence of the balances and relationships and consequently has no relationship to the time response of the process or the automation system. The convergence of the models is a matter of seconds for design but increases to minutes for Real Time Optimization (RTO) depending upon model scope and computing power.

Steady-state models can be used to find the process gains that cause changes in the composition, density, pH, and temperature of process output streams for changes in the process input flows. However, levels are fixed and pressures do not take into account the effect of changing pressure drops with flow across piping resistances and control valves. Consequently, steady-state models are not used to find the process gains for levels and pressures. Steady-state models can be used as dynamic models of composition and temperature by the addition of a dead time and filter blocks on the process outputs for synchronization and correction with analyzer measurements. The dead time block simulates the total loop dead time and the filter blocks simulate the primary and secondary time constants in the loop.

> ***Steady-state models developed for the design of continuous processes can be used for inferential measurements by the addition of dead time and time constants for synchronization and correction with analyzers. This synchronization and correction is similar to what is done for an LDE but with all of the complexities of first principle process relationships built-in.***

Whereas first principle steady-state models zero out the ODE derivative, first principle dynamic models integrate the ODE derivative. The differential equations are largely the same except that crystallization, evaporation, heat transfer, mass transfer, and reaction rates are used instead of equilibrium relationships to provide the dynamics between different operating conditions. The use of reaction kinetics and driving forces for heat and mass transfer rather than equilibriums is particularly important to show the process response during abnormal operation and startup.

The differences in temperature across a surface with heat transfer coefficients and surface area are used as driving forces for heat transfer rates. Similarly, differences in composition across a surface with mass transfer coefficients and surface area are driving forces for mass transfer rates. For modeling dissolved oxygen, the bubble surface area and the composition inside and outside the bubble are used in the driving force calculation for mass transfer.

Dynamic first principle models should include the dynamics of the automation system as well as those of the process as shown in Figure 2.7. Valve and variable speed drive (VSD) models should include the installed characteristic, resolution and sensitivity limits, deadband, dead time, and a velocity limited exponential response. Measurement models should include transportation delays, sensor lag and delays, signal filters, transmitter damping, resolution and sensitivity limits, and discontinuous signal update delays. For analyzers, the measurement models should include sample transportation delays, cycle time, and multiplex time. Wireless device measurement models should include the update rate and trigger level.

Dynamic first principle models must include the automation system dynamics.

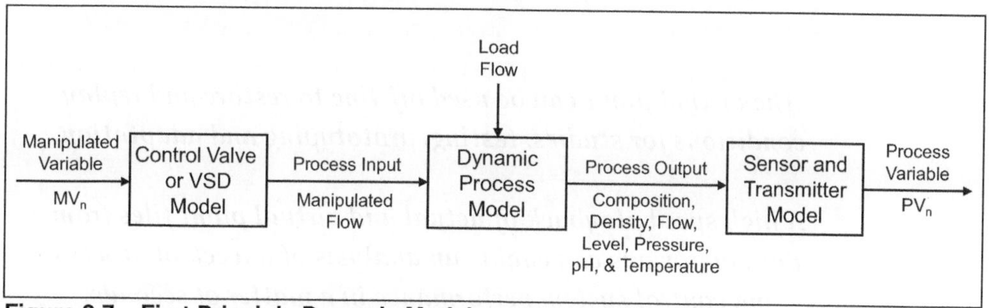

Figure 2.7 – First Principle Dynamic Model

The most effective use of a first principle dynamic model is by means of a virtual plant (VP) whose key feature is the ability to use the actual configuration, historian, displays, and advanced control toolset of the real plant without translation, emulation, special interfaces, or custom modifications. The configuration data base from the real plant can be exported and then imported and downloaded into a personal computer or a control system computer just as if it was an actual hardware controller. Files for operator graphics, process history charts, and data history from the real plant can also be copied to the computer for the VP so that the user has the entire control system of the real plant on a computer.

The virtual plant offers the most effective use of dynamic first principle models by enabling the use of the actual DCS configuration eliminating the need to emulate controls or interfaces.

The virtual plant has the ability to simultaneously stop and start the execution and restore and replay simulated conditions of all the control and the simulation modules. Now emerging is the ability to replay actual or virtual plant data history files at high speeds for adaptation and testing of the process and control modules without a connection to the actual plant. A week of data can be replayed in seconds. This method was extensively used by DCS developers to test the ability of an auto tuner and adaptive controller to identify process dynamics. This capability is well known but will not be commercially offered until customers and third party suppliers of simulation services recognize and demand the incredible power of high speed replay of data history files.

The virtual plant can be used off-line to restore and replay conditions for studies, testing, prototyping and adaptation.

A high speed playback of actual and virtual plant files from data historians can enable an analysis of a week of process and control system performance in a matter of seconds.

The primary PID setpoints of the actual plant are used in the virtual plant. If the virtual plant has the same dynamics as the actual plant, the PID tuning settings in both plants can be the same when the virtual plant is running real time in sync with the actual plant. Since the primary PID controllers are keeping the process variables in both plants at the same setpoint, errors in model parameters show up as differences in the primary PID outputs that manipulate a secondary PID setpoint, valve position, or VSD speed.

Figure 2.8 shows the setup of a virtual plant and the use of model predictive control (MPC) for adaptation of process parameters and optimization of bioreactor profiles. The virtual plant synchronizes with the actual plant at the start of the batch by the initialization of process conditions and PID outputs in the virtual plant. The virtual plant provides inferential measurements of cell growth rate and product formation rate when running real time in synch with the actual plant. The MPC optimizes these rates by keeping them as close to the maximum possible rate captured online just before the inflection point in an uncontrolled profile. The MPC is switched from manual to automatic at the inflection point. The virtual plant and MPC enable more repeatable profiles

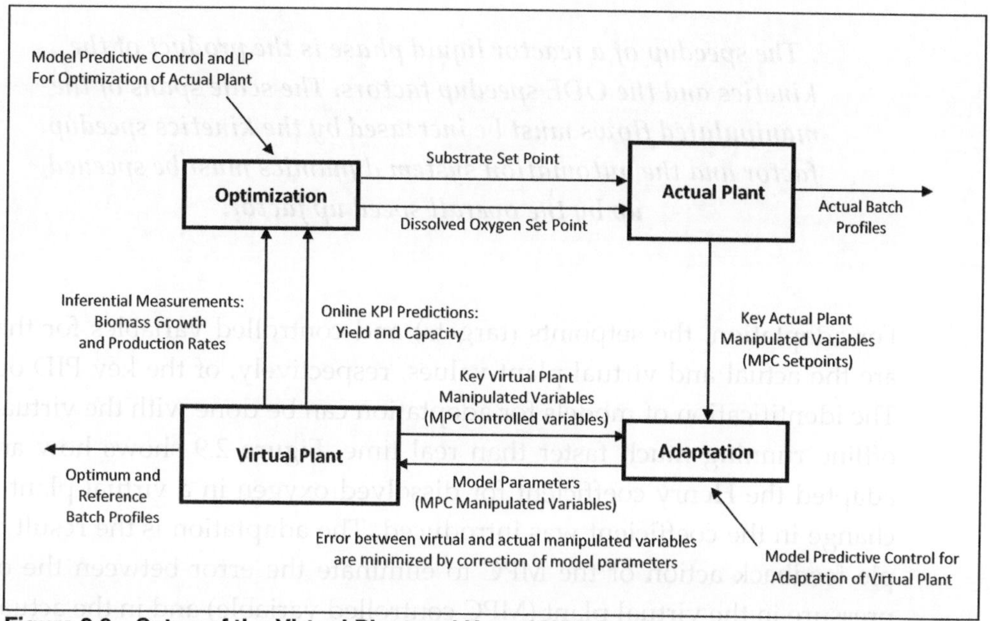

Figure 2.8 – Setup of the Virtual Plant and Use of MPC for Adaptation and Optimization of Bioreactor

and prediction of key performance indicators (KPI) such as yield and capacity than do traditional batch control systems.

For the development and prototyping of new control systems and the analysis of "what-if" scenarios, the virtual plant can be run at about 500 times real time so that a 10-day batch is complete within about half an hour. To see what the batch would be like in the future including the future value of inferential measurements and key performance indicators, a virtual plant running in synch with the actual plant can be switched to run much faster than real time. The real time speedup is the kinetics speedup factor multiplied by the ODE speedup factor for the material, energy, and component balances. The scale ranges of manipulated flows must be scaled up by the kinetic speedup factor. The dynamics of the automation system must be speeded up by the overall speedup factor to keep the PID tuning the same as in the actual plant. The speedup of gas phases is generally not possible due to the small residence time and the integration step size required. A calculation of the blend of results from the liquid phase dynamics and gas feeds can be used to compute the corresponding gas phase concentrations. The ODE for the gas phase mass is still used for the dynamic mass balance used in the pressure model.

The speedup of a reactor liquid phase is the product of the kinetics and the ODE speedup factors. The scale spans of the manipulated flows must be increased by the kinetics speedup factor and the automation system dynamics must be speeded up by the overall speed up factor.

For adaptation, the setpoints (targets) and controlled variables for the MPC are the actual and virtual plant values, respectively, of the key PID outputs. The identification of models for adaptation can be done with the virtual plant offline running much faster than real time. Figure 2.9 shows how an MPC adapted the Henry coefficient for dissolved oxygen in a virtual plant after a change in the coefficient was introduced. The adaptation is the result of simple feedback action of the MPC to eliminate the error between the overlay pressure in the virtual plant (MPC controlled variable) and in the actual plant (MPC setpoint) by manipulation of the Henry coefficient. The overlay pressure was manipulated for dissolved oxygen (DO) control in a cascade control system where the DO PID output was split ranged to a pressure PID setpoint.

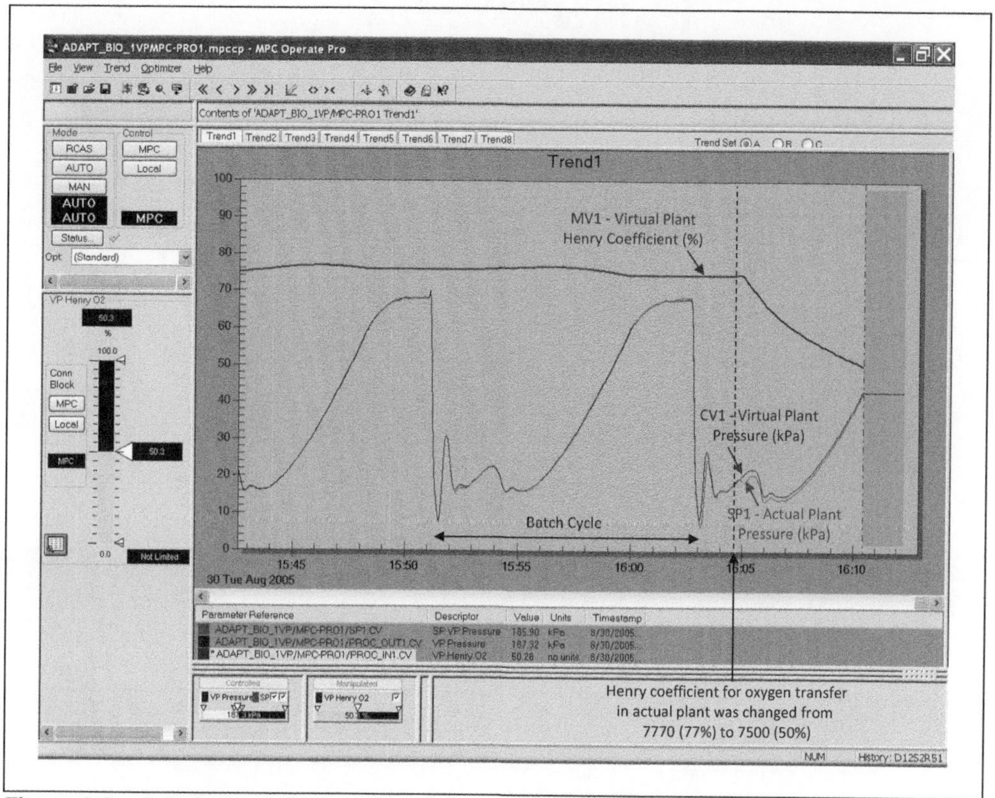

Figure 2.9 – Adaptation of Henry Coefficient in Virtual Plant by MPC Eliminating Error between Virtual Plant and Actual Plant Overlay Pressure in Bioreactor

> *A virtual plant running in synch with an actual plant can be adapted online by the simple addition of a Model Predictive Control or Valve Position Control to adjust model parameters to equalize the corresponding PID outputs in the virtual plant and the actual plant.*

3
Reactor Characterization

3 Reactor Characterization

Reactors processing liquid reactants can be categorized as continuous stirred tank reactors (CSTR), pure batch, fed-batch, and plug flow. In pure batch, all feeds are sequenced with flow setpoints fixed by batch logic. In fed-batch, feed flows are ratioed and manipulated for composition control. In continuous reactors, there is a continuous liquid discharge flow. Often the control systems for a fed-batch reactor can be the same as for a continuous reactor except there is no level control. In batch operations, the liquid discharge flow has an on-off valve that is opened to empty the reactor. A level controller is used in the CSTR to maintain a level that provides enough reaction time for a given production rate. In pure batch and fed-batch reactors, the liquid discharge valve is closed until the batch is ready for transfer. A gas product or byproduct may be continuously generated during the batch and accumulated in an overhead system often as a condensable vapor. A liquid and/or solids phase reaction without a continuous liquid and/or solids discharge flow is the distinguishing characteristic of pure batch and fed-batch. The implications as to dynamics, yield, and inventory control are significant.

Table 3.1 summarizes how the dynamic response, residence time, reaction rate, and controlled variables vary with reactor type as discussed in the following sections. The reactor types considered are pure batch, fed-batch, plug flow, gas flow, and continuous stirred reactors (CSTR).

Table 3.1 – Reactor Type Dynamics and Control

(* additional control besides temperature and pressure control)

Reactor Type	Dynamic Response	Residence Time Distribution	Reaction Rate	Additional Control*
CSTR	Near-Regulating Runaway	Wide	Moderate	Level Composition
Pure Batch	Integrating Runaway	Very Tight	Slow	None
Fed-Batch	Integrating Runaway	Very Tight	Slow	Time Profile
Plug Flow	Self-Regulating Runaway	Tight	Fast	Length Profile
Gas	Self-Regulating Runaway	Tight	Very Fast	Composition

Gas flow reactors are almost always operated in a continuous mode. The reactants and products are all in the gas phase. There may be a fluidized catalyst bed. There is negligible back mixing resulting in plug flow. What distinguishes a gas flow reactor from a plug flow liquid reactor is a gas reaction residence time (e.g., 2 seconds) that is orders of magnitude smaller than the liquid reaction residence time (e.g., 10 minutes). Consequently, the temperature profile control strategy in a liquid plug flow reactor to monitor and set the progress of the reaction with respect to length is not used in gas flow reactors because the reaction goes so quickly toward completion. An average of a cross-sectional traverse of the gas flow reactor with a hot spot override is used for fluidized bed temperature control rather than temperature profile control.

3-1. Classification of Dynamic Responses

Reactor type and reaction kinetics determine the dynamic responses that are best distinguished when the PID is put in manual and a step change is made in the PID output. For the purposes of identification and characterization of the open loop response, there are no process disturbances or effects from other loops so that the response observed is totally the result of the change in the PID output. The different types of dynamic responses were described in Chapter 1 but are summarized here to facilitate the discussion of how equipment design affects these responses.

If the response settles out at a new value (steady state), the process is termed self-regulating. Such processes are further categorized according to the ratio of the primary time constant to the total loop dead time as being dead time dominant, moderate self-regulating and near-integrating. While any process with a ratio less than 1 may be considered dead time dominant, so far as PID tuning and control loop performance are concerned, a process is not considered to be dead time dominant until the ratio is less than ¼. If the ratio is between ¼ and 4, the process is considered to have a moderate self-regulating response. Most of the literature is centered on tuning and improving the control system performance for this type of response. If the ratio is greater than 4, the process is considered to have a near-integrating response. This last classification is critical if lambda tuning is used to get the tuning settings needed to stop (arrest) an excursion from a load disturbance that is not slowing down.

To enable proper PID tuning and loop analysis, self-regulating processes whose open loop response would settle out at a new steady state are classified as dead time dominant, moderate self-regulating, or near-integrating based on the primary time constant to dead time ratio.

If the open loop response continues to ramp and doesn't decelerate or accelerate, the process is classified as integrating. The process will ramp for true integrating processes until a physical limit is reached or a Safety Instrumented System (SIS) is activated (e.g., relief valve blows or equipment shutdown) if the PID is left in manual. The near-integrating process looks like a ramp in the time frame of the PID response (e.g., two dead times) but will eventually decelerate. The tuning rules are the same for near-integrating and true integrating processes.

A process is classified as true integrating if the open loop response continues to ramp until a physical limit is reached or an SIS is activated. A process is near-integrating if the open loop response appears to be a ramp just in the time frame of the major change in PID output.

If the response accelerates with a continually increasing rate of change, the process is classified as runaway. The system is intentionally designed so that acceleration does not start until the excursion is exceptionally far from setpoint. In order to prevent environmental releases, shutdowns, equipment damage, and hazardous operation from a runaway reaction, these processes are not allowed to stay in manual long enough to see acceleration.

Runaway processes are treated as near-integrating. Since these processes typically have a high PID gain to prevent large excursions, the PID can stay in automatic and the step change in PID output from a setpoint change used to identify the process dynamics if the PID Structure is PI on error and D on PV. Just the dead time and initial ramp rate needs to be identified. The setpoint change can be quite small since the PID gain is so high.

Acceleration in the open loop response of a runaway process is not allowed to occur. The process response is treated as near-integrating and the dynamics for PID tuning are identified from a small setpoint change.

An initial response of any of these processes in the opposite direction of the eventual response is termed an "inverse response." Temperature control by manipulation of cold feed can exhibit an inverse response particularly for slow reactions. An inverse response necessitates reducing the PID gain to reduce the initial reaction of the PID that is in the wrong direction.

Reactor temperature control systems with slow kinetics that manipulate feeds with a temperature significantly different than the operating temperature, can exhibit an inverse response where the initial response to a feed change is in the opposite direction of the final change in temperature.

A continuous stirred tank reactor (CSTR) has a slow near-integrating response due to the residence time becoming a large primary time constant from good back mixing. A pure batch and fed-batch reactor have a slow true integrating response. A continuous plug flow reactor has a moderate self-regulating response for temperature control by manipulation of coolant due to the thermal time constant introduced by heat transfer surfaces. Inline polymerization reactors, static mixers, extruders, and gas reactors (e.g., fluidized bed reactors) are considered to be plug flow reactors because there is negligible back mixing (negligible axial mixing).

In gas reactors feeds can be manipulated for temperature control because the reaction kinetics are so fast that there is little or no inverse response at the points chosen for temperature control. For these systems, coolant coils are switched in and out of service to set cooling capacity and hence production rate. The process response is still moderate self-regulating when temperature manipulates feed due to the thermowell time constant. All of these reactors can develop a runaway response when the increase in reaction heat release with temperature exceeds the cooling capability.

Temperature control of a continuous stirred tank reactor (CSTR) has a slow near-integrating response, a batch reactor has a slow true integrating response, and a continuous plug flow reactor has a fast moderate self-regulating response. All of these reactors can develop a runaway response if the increase in exothermic reaction heat release rate with temperature exceeds the cooling rate capability.

The use of an analyzer with a large sample, cycle, and/or multiplex time almost guarantees dead time dominance (total loop dead time larger than open loop time constant) for composition control of continuous, plug flow, and gas reactors. Inferential measurements, online analyzers, and fast at-line analyzers are used to minimize the introduction of excessive dead time into a concentration control loop.

The use of slow at-line analyzers for concentration control of continuous reactors can create a dead time dominant loop.

The introduction of a slow temperature sensor or slow analyzer can be disastrous for reactors that can develop a runaway response. Section 1.3.8 discussed the window of allowable PID gains and Equation 1.10 detailed the requirement that the dead time or secondary time constant must be much smaller than the primary positive feedback time constant otherwise the window closes causing instability for all tuning settings. Fouling of heat transfer surfaces for coolant and for thermowells adds an excessive secondary time constant. Slow analyzers add excessive dead time.

The fouling of heat transfer surfaces or a long sample transportation time or analyzer cycle time or multiplex time can cause a reactor with a runaway response to be unstable for all tuning settings.

Sometimes fed-batch is called semi-batch or semi-continuous. However, there is an important distinction between fed-batch and continuous reactors; in that in fed-batch as in pure batch reactors there is no liquid discharge flow until the end of the batch. The lack of a liquid discharge flow is an integrating response and a rise in level with reactant addition. The increase in level from start to finish of the batch and the lack of self-regulation as a result of the closed discharge valve affects the dynamic response of temperature and composition requiring tuning methods and settings not commonly discussed in the control literature. For example, a temperature controller gain of 50 or more is possible for fed-batch reactors.

3-2. Residence Time

Residence time is critical for reactors; it determines the amount of time available for reaction by virtue of the reactants being in contact with each other. The reaction time must be less than the residence time for the reaction to go to completion. If the residence time is too long, undesirable reactions and product degradation may occur. For maximum yield, the residence time should be slightly larger than the reaction time with a tight residence time distribution (statistical plot of each component population versus residence time should be a narrow peak). For maximum capacity and minimum capital cost, the residence time should be as small and the distribution as tight as possible since residence time is proportional to cycle time for batch and volume for continuous operations.

For maximum process efficiency and capacity and minimum capital cost, the residence time should only be slightly longer than reaction time, with an extremely tight residence time distribution.

For batch operations the batch cycle time is extended past the feed phase via a hold phase to ensure the reaction is complete. The residence time is the average volume divided by average feed rate during the feed phase plus the time in the hold phase. For pure batch operations where all of the feeds are added quickly and simultaneously, all of the components are in the reactor for about the same length of time making the residence time distribution tight.

In fed-batch operations the feeds are added via a cascade control and ratio control system. In most cases, a primary loop for concentration control is manipulating the setpoint of a secondary loop for the leader feed flow. The other feed flows have setpoints ratioed to the leader feed flow setpoint. The rate of feed addition for fed-batch operation may start out fast but is gradually reduced as the reaction nears completion exhibiting a nonlinear profile consistent with reaction rate. In a pure batch operation, the feed rates stay fast and fixed unless sequenced by batch logic to lower rates near the end of the batch in anticipation of reaching the endpoint. The difference from one batch to another is not seen in the concentration profile but in the profile of the manipulated feed flows if the cascade control system is doing a good job. The residence time of components added early in the feed phase is longer adding a slight degree of uncertainty in the residence time distribution. Batch operations are considered here to be well-mixed so that turnover time is negligible in terms of residence time. The downside of batch operations is a greater equipment cost to achieve the same capacity as continuous operations.

Batch operations ensure a residence time greater than the reaction time and an extremely tight residence time distribution but offer less capacity than continuous operations.

The overall residence time for continuous operations is the current operating volume divided by the total flow rate passing through the volume. Some tuning methods for surge tank level control inadvertently use a definition of residence time as the maximum volume divided by the maximum change in flow set by equipment size, level connection location, and PID scale ranges. Here we are using a definition of residence time that changes with operating conditions as well as with equipment design. At higher production rates, the residence time gets shorter for reactions unless the level setpoint is increased accordingly.

The residence time for continuous operations is the operating volume divided by the total flow rate.

A plug flow reactor has no back mixing (axial mixing). Polymer plug flow reactors use inline equipment such as pipes or tube(s). In static mixers baffles create some radial mixing. In extruders, a rotating screw pushes a viscous liquid and solids to the exit. Since there is no back mixing, all of the components in the reaction mass exit at the same time, a time that is equal to the transportation delay. All of the residence time becomes a process dead time. A plug flow reactor is similar to a fed-batch reactor in that there is generally no level control and the residence time distribution is extremely tight. If you consider a subsection of the reaction mass as moving from the entrance to the exit, there is no discharge from this volume until the subsection approaches the exit.

The residence time in plug flow reactors is a transportation delay and the residence time distribution is tight.

Polymerization processes often use pure batch or fed-batch and plug flow reactors operating with high viscosities. For highly viscous products the residence time distribution is not as tight. Product can accumulate on the vessel walls in batch reactors. The velocity near the pipe wall in plug flow reactors is slower from frictional drag causing the product near the wall to arrive later. The increase in time available for reaction increases the viscosity of the fluid near the wall aggravating the problem.

The residence time distribution of polymerization reactors is not as tight as with other plug flow reactors operating at lower viscosities due to product accumulation and friction drag along the walls.

Gas phase reactors are usually continuous plug flow reactors. Many use fluidized catalyst beds. There is little to no back mixing except from turbulence. Consequently, the residence time distribution is tight for a uniform catalyst and flow distribution. The dynamic response to changes in reactant feed is a transportation delay similar to the plug flow liquid reactor except the reaction times are faster and the residence times are much shorter from a much higher velocity. Channeling of gas flows through the catalyst bed can, however, result in some gases arriving at the outlet sooner reducing residence time and conversion for some portions of the exiting gas product.

The residence time of gas phase reactors is short and the residence time distribution is tight where there is an even flow distribution.

In a continuous stirred tank reactor (CSTR) some of the reactants appear in the discharge within one turnover time, which in general is less than 1/20 of the residence time in a well-mixed vessel. The residence time becomes a large primary time constant causing a near-integrating response. The amount of time spent by a given component in the reactor varies resulting in a loose residence time distribution and a loss in yield in the CSTR. Reactor designs with a liquid height about equal to the diameter, baffles to prevent swirling, an axial agitation pattern that breaks the surface, and a dip tube design to bring reactants into the eye of the impeller and away from discharge nozzles to avoid short circuiting have a tighter residence distribution. Several CSTR in series or a reactant recovery and recycle system can be used to improve the yield of the production unit particularly for slow reactions and high purity requirements. However the use of a recycle system creates an integrating response and increases the propensity for buildup of inerts and impurities requiring the optimization of recycle flow and a purge flow for the best yield.

The residence time in continuous stirred reactors becomes a primary time constant and the residence time distribution is loose but can be tightened by good equipment design.

3-3. Control Loops

Nearly all reactors have temperature and pressure control since these affect reaction rate. Concentration control is desirable but depends upon the feasibility, accuracy, and reliability of analyzers and inferential measurements. For at-line analyzers, there is the additional consideration of the dead time added to the loop. To date offline analyzers have only been used to correct inferential measurements or to make manual adjustments. The use of an enhanced PID opens the door to the possibility of using off-line analyzers directly for control. A large and variable analysis time does not cause instability in a loop with an enhanced PID. The time between offline analysis results still needs to be made as short as possible for the PID to correct for disturbances since the PID only makes a correction when there is an update.

Nearly all reactors need temperature and pressure control and would benefit from concentration control.

There is generally no level control except perhaps in terms of a high level override or high level shutdown of feeds in pure batch and fed-batch reactors. In batch reactors, the reactants are fed sequentially and shutoff when charge tank weight measurements or flow totals indicate that the total amount charged is complete. On-off isolation valves are used to start and stop feeds. Flow controllers and control valves are sometimes used to prevent vapor system overload and provide a repeatable feed cycle time. In fed-batch operation, the reactants are fed simultaneously at a rate determined by a control system. As a minimum there are flow controllers for each reactant.

Many of the same control systems used for continuous reactors are applicable to fed-batch, except that there is typically no level and residence time control. The literature does not lead one to understand that the pressure, temperature, and composition controllers used in continuous reactors to manipulate reactant flow setpoints can be used for fed-batch. The use of these cascade control systems in fed-batch may even be more important because of the opportunity of these systems to provide batch profile control.

Many of the control systems used for continuous reactor control can be used for fed-batch control. The exception is that level control is not needed except possibly as an override.

Relatively fast and tight temperature control is possible for gas reactors by the manipulation of reactant feed rate due to the lack of inverse response and the small process dead time. The process dead time from the transportation delay of reactants is small compared to the time constants from catalyst heat capacity and thermowell design.

> ***Relatively fast and tight temperature control is possible for gas reactors due to a small process dead time by the manipulation of reactant feed rate rather than coolant.***

There is an optimum profile of temperature, physical properties, and composition with respect to length for a plug flow reactor. For a batch vessel there is an optimum profile with respect to batch time. For extruders, the decrease in temperature and the increase of viscosity with length can be optimized. Optimum profiles may also be applicable with respect to startup time and transition time of continuous reactors. There are more opportunities for profile control and optimization than presently realized in the process industry.

> ***Optimum profiles exist with respect to time for batch reactors, during the startup and transition of continuous reactors and with respect to length for plug flow reactors.***

Since intermediate and final product composition generally go in one direction only with time for nonreversible reactions, the control of an absolute value whose setpoint varies with time can result in a unidirectional response preventing the use of integral action, as discussed in Chapter 1. The use of the slope of a composition profile as the controlled variable enables a bidirectional response allowing the use of a standard PID structure and tuning. The implementation of this strategy will be extensively discussed in Section 5.8 on biological fed-batch profile control. This strategy can be applied to intermediate and final products and byproducts. For example, bioreactor cell growth rate is controlled early in the batch and lactate formation and product formation rate is controlled late in the batch.

> ***The slopes of key compositions with respect to time should be controlled rather than a concentration value whose setpoint is varied with time.***

3-4. Pure Batch versus Fed-Batch versus Continuous

In pure batch operations, changes in feed conditions, disturbances, or demand show up as changes in the profile of key process variables. For a bioreactor, variability in the seed culture, indeterminacy of cell response and changes in the demand for air, substrate (e.g., glucose), and reagent cause the variation in the batch profile for dissolved oxygen and substrate concentration and pH. Process engineers can be ingenious at changing the timing and flow rate of air, substrate, and reagent via sequential scheduling as shown in Figure 3.1a. However process knowledge is far from perfect and unknowns can change the profiles of the key process variables.

The sequential scheduling of batch feeds in pure batch operations presumes complete knowledge and a lack of variability in raw materials and batch behavior.

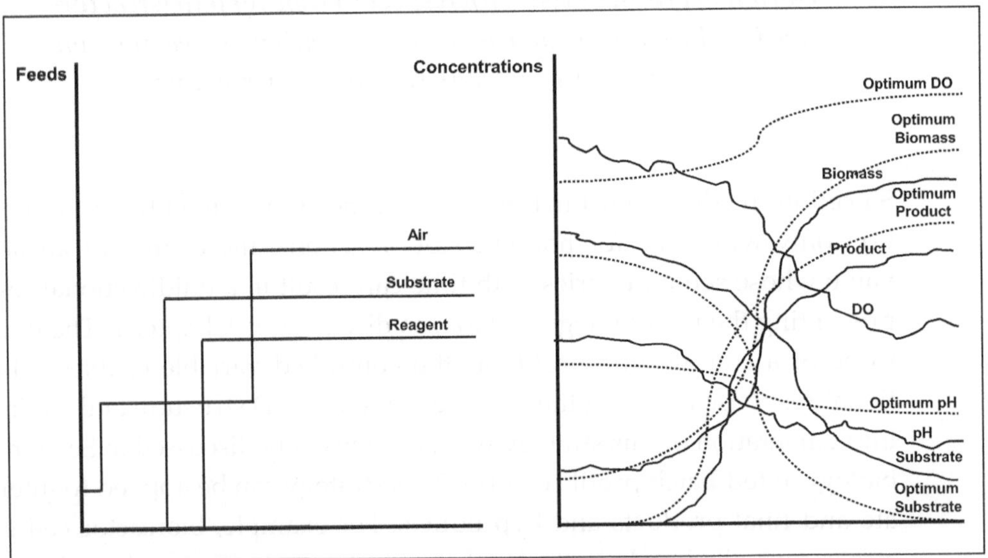

Figure 3.1a – In Pure Batch Operations, the Flows are Sequentially Scheduled Leaving Key Process Variables Susceptible to Operating Conditions, Uncertainties, and Disturbances

The fed-batch control system in Figure 3.1b enables the dissolved oxygen and substrate concentration and pH to be more constant throughout the batch. The variability in these controlled variables is transferred to the manipulated air, substrate, and reagent flows. At first this can be disquieting because operations personnel have now lost the ability to force these flows to follow a fixed pattern.

The definition of a controlled variable and a manipulated variable is based on the latter serving the needs of the former and not vice versa. The final resting value of manipulated variables can be optimized by slowly adjusting controlled variables. This is in fact the strategy of valve position controllers that seek to increase production rate by keeping a coolant valve at the maximum controllable position by manipulating a feed flow. One cannot keep constant both the controlled variable and the manipulated variable.

The feedback control key process variables by the manipulation of batch feeds in fed-batch will automatically correct for unknowns and disturbances.

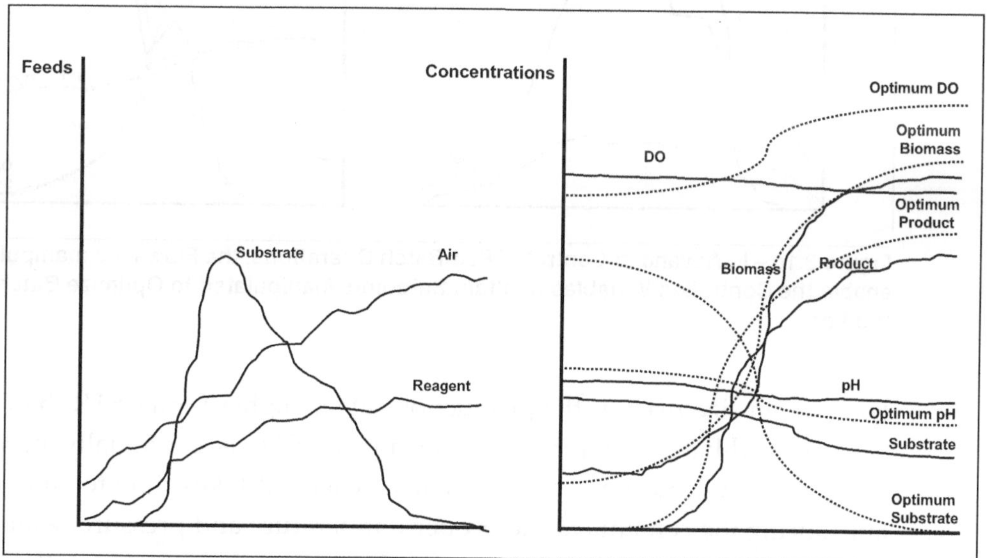

Figure 3.1b – In Regulatory Control of Fed-Batch Operations, the Flows are manipulated to keep Key Controlled Variables at a Fixed Setpoint

The optimized fed-batch control system in Figure 3.1c makes adds a higher mode of control. A model predictive controller whose controlled variables are the slope of the biomass concentration profile (cell growth rate) and the slope of the product concentration profile (product formation rate) manipulates the dissolved oxygen and substrate concentration setpoints. The result is a batch profile of these higher controlled variables that more closely matches their optimums.

> ***The addition of an upper mode of control by the addition of changes of compositions and temperature with respect to time as controlled variables can optimize batch profiles.***

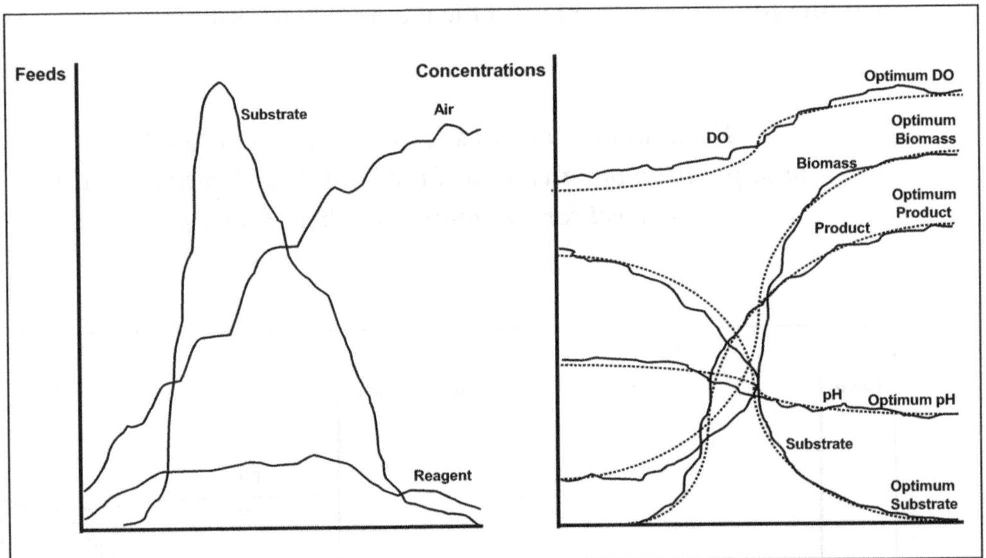

Figure 3.1c – In Advanced Control of Fed-Batch Operations, the Flows are manipulated to enable the Controlled Variables to attain Setpoints Manipulated to Optimize Batch Profiles

For a given size reactor, the production rate is higher for a CSTR than a pure batch or fed-batch reactor. After startup a CSTR is continually producing whereas with a pure batch or fed-batch reactor it takes time to charge reactants, bring the contents to the proper temperature and pressure, wait for the reaction to complete, and then empty the reactor. In some cases the reactor may need to wait on shared resources such as feed weigh tanks or on operator actions or lab results from sequential actions.

Continuous processes cost less in terms of investment and utilities for a given capacity but generally require greater process research, development, and design. Mature high capacity products (e.g., oil, gas, and petrochemicals) tend to use continuous reactors whereas new high value added products (e.g., specialty chemical and biological) primarily use pure batch and fed-batch reactors. As volumes increase and the profit margins decrease (the product becomes a commodity), there is increased emphasis on developing a continuous process to provide a higher capacity and lower operating cost.

Candidates for continuous reactors are products with a low profit margin, a high volume requirement, fast reaction rates, minimal adverse reactions, preventable buildup of inhibitors and inactive components, inerts, and extensive R&D history leading to a deep fundamental knowledge of kinetics. Oil, gas, chemical intermediates, petrochemicals, and commodity chemicals use continuous reactors. Extensive integration of unit operations for energy recovery and recycle of materials offers complex opportunities for optimization.

Continuous reactors offer more capacity for a given size or a lower capital cost and greater energy efficiency for a given capacity but require greater knowledge and effort in research, development, and design.

The fastest and simplest implementation is pure batch with quantities charged sequentially mimicking lab experiments much like ingredients in a recipe. As knowledge is gained batch reactors can move to become fed-batch reactors and eventually continuous reactors if there is enough demand and if reaction chemistry permits a variable residence time.

In contrast to continuous reactors candidates for batch reactors are products with a high profit margin, low volume requirement, slow reaction rate, significant side effects, and minimal knowledge of kinetics. Specialty chemicals, food, beverages, and especially biopharmaceuticals are produced by batch reactors. For new biopharmaceuticals with the patent expiration clock ticking and extraordinarily high prices (e.g., > $1000 per gram), the time to market supersedes any consideration of process efficiency or capital cost. These processes have extremely slow kinetics, requiring days to weeks for the cells to produce product.

> ***Batch operations require less process knowledge and customization of equipment and offer a fixed and tight residence time distribution enabling a more certain result and a faster project when a new product is commercialized.***

The buildup of dead and mutated cells, toxins, inhibitors, viruses, and bacteria can be prevented in batch processes by emptying, cleaning, and sterilizing the equipment and piping after each batch. The adverse results of contamination to food, beverages, and drugs can be extreme in terms of the loss of capacity, cost of retrieval, and most of all the consequences to the customers.

> ***Batch processes enable the periodic emptying, cleaning and sterilization of equipment and piping to prevent the buildup of contaminants in products that end up in a person or animal.***

Some mature biopharmaceuticals are produced by "perfusion" processes operating in the continuous mode for months at a time with extensive recycle due to the slow reaction rates. Product is removed from the discharge stream and the recycle stream is purified before being returned to the bioreactor. These processes are periodically shut down for total elimination of mutated cells, inhibitors and alien organisms and decontamination.

4
Reaction Kinetics

4
Reaction Kinetics

4-1. Reaction Rate

The yield (amount of product formed per amount of reactant added) and consequently the product quality and capacity depend upon reaction rate. Reaction rate depends upon temperature and composition for all reactions and in addition, upon pressure for gas phase reactions.

4.1.1 Temperature Implications

The reaction rate can increase dramatically with an increase in temperature. Higher temperatures can increase product formation rate but trigger side reactions to form undesirable byproducts, degrade product, and in the case of biological reactions can cause cell death. For biological reactions, during the start of a batch and for subsequent temperature shifts an overshoot of just 0.1 ^{o}C above the optimum setpoint can degrade cell functions. Since temperature is readily measured, a considerable increase in reaction rate is often possible by acquiring knowledge of the optimum temperature setpoint, using accurate temperature sensors, and applying tight temperature control.

> ***Nearly all reaction rates are extremely sensitive to temperature. In general, reactors are run slightly below the absolute optimum because of the dire consequences of higher temperatures.***

4.1.2 Concentration Implications

In addition to temperature and possibly pressure, reaction rates also depend upon the concentrations of various compounds added or formed during the reaction.

The concentration of acids and bases becomes important when acids or bases are added, consumed or produced by reactions. pH affects some chemical and all biological reactions. Mammalian cells are supersensitive to the concentration of hydrogen ions. Cell life and productivity depends upon a narrow window of pH (e.g., 0.1 pH). Additionally the accumulation of sodium bicarbonate used as a basic reagent for mammalian cell cultures can cause a build-up of osmotic pressure (internal cell pressure) in these cells resulting in cell lysis (fracture of cell membranes). For all of these important ionic components pH provides a concentration measurement with extraordinary resolution and rangeability.

pH provides a hydrogen ion concentration measurement of extraordinary resolution, which is extremely important for biological processes due to the extreme sensitivity of cells to pH.

Reaction rate depends upon the concentration of reactants, initiators, promoters, inhibitors, byproducts, waste products, and primary products. In chemical reactions the reactants are chemical compounds. For catalytic reactions, the reaction rate also depends upon the concentration and condition of the catalyst. In polymerization reactions, the reactants are monomers (joined chemical compounds) and the product is a chain of these monomers. In biological reactions, the reactants are oxygen and food sources, such as glucose and glutamine, called substrates. Waste products such as acetate, ammonia, carbon dioxide, glutamate, lactate and phosphate inhibit cell growth rate or product formation rate. Some of these waste products affect cell pH critical for the cell health and growth. Biological products are either held in the cell or secreted. The oldest products, such as alcohol for beverages and fuel, are simple chemical compounds secreted by yeast. New pharmaceutical products are complex proteins found in the body that require the mammalian cell, an incredibly sophisticated reactor, for formation.

Reaction rate depends upon the concentration of reactants, initiators, promoters, inhibiters, byproducts, waste products, and primary products.

To prevent an excess or deficiency of reactants, the reactant concentrations must be in the ratio set by the stoichiometric equation, the fundamental basis for the reaction. Reactant feed flow setpoints are often ratioed to help maintain the right concentration ratio in the reactor. However, errors in flow measurements, stick-slip and backlash in valves, differences in flow loop responses and non-ideal mixing, cause concentrations in the reactor to vary from the ideal ratio.

Often overlooked are the temporary unbalances in the stoichiometric ratios of reactants added resulting in an accumulation of excess or a deficiency of reactants reducing yield.

Chromatographs can be used for reactant and product concentrations in chemical reactions, but the significant resource requirement often discourages small companies and "small products" industry (manufacturers with a small market for products) from making the investment in the capital and expertise required for the design, installation, and maintenance of the analyzer and the sample system. Plants with high capacity products such as petrochemicals and intermediates greatly depend upon the added value of chromatographs because even a fractional percent increase in production rate is worth millions of dollars. There are often parallel trains of reactors that can share an at-line analyzer (e.g., chromatograph), thereby reducing capital and maintenance costs. For polymerization reactions gel chromatographs are used to determine molecular weight distribution for yield prediction and control.

Unless such analyzers are part of the plant culture, their value may be discounted. Without analytical measurements, the possibility and benefits of doing something better in terms of process control and optimization may not be recognized.

The benefits of at-line chromatographs as a general solution for definitive and accurate concentration measurements are generally underestimated.

In some cases, viscosity meters can provide inferential measurements of average molecular weight. Polymer quality also depends upon characteristics other than composition, such as impact resistance, tensile strength, color, and optical clarity that require special lab tests. The inferential measurements described in Chapter 2 can enable fast online concentration control by eliminating the dead time from at-line and offline analyzers.

Biological reactions utilize an unusually large variety of composition measurements including potentiometric and amperiometric electrodes for substrates and waste products (e.g., pH, dissolved oxygen, glucose, dissolved carbon dioxide, and specific ion probes), turbidity for biomass, dielectric spectroscopy (e.g., capacitance probes) for viable versus lysed cells, digital imagery for viable cell size and concentration, near infrared (NIR) spectroscopy for substrate and product concentration, and mass spectrometers for off-gas concentrations. The analysis of the folding of complex proteins currently requires special tests that can take days.

Biological reactions use a wide variety of online, at-line, and offline analyzers.

Inventory control by the use of simple pressure or level control can be used to help prevent an excess accumulation of a reactant. A pressure controller manipulating a gas reactant feed rate for a liquid product and a level controller manipulating a liquid reactant for a gas product will automatically add reactant at the rate at which the reactant is consumed and then shutoff the feed when the reaction stops. This relationship holds true for pure batch, fed-batch, or continuous reactors. A purge from the opposite phase of the product is necessary to prevent the accumulation of inerts or non-reactants that reduces the volume of reactants. For a gas product, a liquid purge is needed and for a liquid product, a gas purge is needed.

4-2. Reaction Time

Neutralization acid-base reactions are essentially instantaneous and occur as fast as mixing allows. Most other reactions have a measurable reaction rate. The amount of time that reactants stay in contact is called the residence time in continuous operations and is simply the reactant phase volume divided by the total volumetric phase flow rate. If the phase volume does not change, the residence time changes with production rate for continuous operations. If significant solids buildup on the wall in polymer reactions or inert accumulation occurs, the volume of reactants and consequently the residence time are reduced. If this occurs level setpoints can be increased to provide a larger volume to help keep the residence time constant. For batch operations, the batch phase and cycle time must be long enough to provide the reaction time needed.

As plant capacity is pushed by increasing feed rate or decreasing batch cycle time, care must be taken that the reactor provides enough reaction time.

4.2.1 Batch Operations

For fed-batch and pure batch operations, a batch cycle time is used instead of the residence time for the liquid phase. If the time is too short, the reaction will not be complete. If the time is too long, side reactions producing waste products may develop and reverse reactions may occur.

In batch operations, the batch cycle time is often longer than necessary for completion of the reaction. The primary reasons for this practice are the lack of online composition measurements, a conservative approach, insufficient research, no downsides other than capacity (no product degradation), and the simplicity of a fixed batch cycle time. Consequently batch reactions are candidates more for capacity improvement than for yield improvement. Batch cycle time improvements are often easy to implement and offer an increase in capacity that is 10 times greater than typical improvements in continuous processes (e.g., capacity improvements of 20% for batch versus 2% for continuous).

Batch operations generally have a greater potential for capacity improvement than continuous operations.

4.2.2 Continuous Operations

In back mixed continuous operations, some reactants end up in the exit flow because a non-zero reactant concentration is necessary for the reaction to proceed and a well-mixed volume means that some of the reactant is in the discharge flow. For continuous reactions requiring precise residence time control, a plug flow reactor is used that by definition has no axial or back mixing. The lack of back mixing means that the time the reactant takes to go from the entrance to the exit of the reactor is constant. Pipeline reactors, static mixer, gas reactors, and extruders are examples of plug flow reactors. Eliminating the variation in residence time for reactants provides a more repeatable reaction result and a greater yield. However, for concentration control by manipulation of feeds, the residence time is a transportation delay rather than a primary process time constant. Transportation delay adds to the total loop dead time that delays the recognition and correction of a disturbance. In contrast, a process time constant slows down disturbances giving the controller time for correction and acts as a filter to attenuate oscillations from reactant pressure fluctuations or valve stick-slip and from noise resulting from non-ideal mixing. The peak error and integrated error for disturbances are inversely proportional to the PID gain as detailed in Section 1.4. The maximum PID gain is proportional to the primary time constant to dead time ratio as extensively discussed in Section 1.3, and is limited by the amount of noise.

Plug flow reactors are used to provide a more repeatable reaction and greater yield.

If a pipe is used there is little radial mixing. If a static mixer (baffled section of pipe) is used there is considerably more radial mixing. High viscosity reduces radial mixing. The cross sectional variation in reactant concentration increases as radial mixing decreases. If the plug flow reactor is operated with no heat transfer, there is also very little temperature variation in the cross section except with high viscosity fluids due to less radial mixing and more heating along the wall from shear. If temperature control is used to vary the heat transfer, there is a temperature gradient from the wall to the center of the pipe or static mixer. There is an axial temperature and concentration gradient along the length of the reactor as the reactants react in the path from entrance to exit.

The temperature profile may be controlled by manipulation of heat transfer fluid to zones of the reactor.

An extruder is essentially a plug flow reactor with a moving radial agitator (screw) where the polymer is processed as it moves from entrance to exit. Extruder zones are temperature controlled. The rotation of the screw is a greater source of energy input than the heat from zones and screw speed has a greater impact on product quality than zone temperatures do. There is an optimum specific energy consumption (SEC) (kJ/kg or BTU/lb) for a given residence time. Since the extruder volume is fixed, the residence time is proportional to the feed rate. There are generally efficiency ellipses in a plot of SEC versus feed rate, with the greatest efficiency at the center. There are profiles of physical properties (e.g., viscosity) and product quality parameters along the length from entrance to exit that may not be directly affected by zone temperatures.

Plug flow reactors have axial profiles in temperature and composition. High viscosity fluids (e.g., polymers) have radial profiles as well.

4-3. Chemical Reactions

The stoichiometric equation defines the reaction and provides important knowledge of the proper ratios of reactants. For conceptual purposes a simple reaction of components A and B to form product C is illustrated (Equation 4.1). In this stoichiometric equation "a" moles of component A react with "b" moles of component B to form "c" moles of product C. To get the equivalent reactant or product mass (e.g., kg), the moles are multiplied by the molecular weights (MW) of the components.

At excessively high product concentrations or temperatures, a reverse reaction may dominate or completely different reactions may occur where the reactants and product form waste products or undesirable byproducts. Product degradation may also occur at high temperatures. Product qualities such as color may suffer.

The Arrhenius Equation defines the most prevalent relationship between reaction rate and temperature (Equation 4.2). The activation energy determines the severity of the exponential response with temperature. Reaction rates with lower activation energies are more sensitive to temperature. For many common reactions, the activation energy is low enough to cause the reaction rate to approximately double for every 10°C increase in temperature.

> ***Many chemical processes have relatively low activation energies, causing the reaction rate to double for every 10°C increase in temperature.***

The reaction rate is the Arrhenius Equation result multiplied by reactant concentrations with exponents that determine the order of the reaction (Equation 4.3). A reaction is first order with respect to reactant concentration A (C_A) if the corresponding "α" exponent is unity, making the reaction rate proportional to reactant concentration. Since the reaction rate depends upon the product of reactant concentrations, the reaction rate decreases as any one of the reactant concentrations decreases. Maintaining the reactants in the proper stoichiometric ratio defined by the coefficients "a" and "b" of the stoichiometric equation is thus critical for maximizing a reaction rate that insures minimal excess reactant left at the end of a batch or exiting a continuous reaction.

To achieve the proper stoichiometric ratio, a reactant flow follows a "leader" reactant flow per the correct ratio. Consider reactant A as the "leader" and reactant B as the "follower". The production rate demand sets the "leader" flow setpoint (reactant A). The "follower" flow setpoint (reactant B) is computed as the reactant A flow setpoint multiplied by the reactant mass flow ratio $R_{B/A}$ (Equation 4a). The mass flow ratio is computed from the stoichiometric equation coefficients and molecular weights for components A and B (Equation 4b). The reactant flow setpoint rather than measurement is used to prevent chasing noise and limit cycles, and adding delays in responding to changes in production rate. The controllers are tuned with as identical as possible closed loop time constants (lambdas) so the flows respond in unison maintaining the proper ratio. However, if there are valve problems, such as plugging or flashing, where the flow measurement deviates significantly from setpoint, the flow measurement rather than the setpoint of the "leader" should be multiplied by the flow ratio to set the setpoint of the "follower" flow controller. Equations 4.1 through 4.4b provide a simplistic view that is useful in showing the first principles for maximizing for reaction rate and maintaining the stoichiometric ratio by process control. These types of equations are also used to define byproduct and reverse reactions. The nomenclature with units is defined at the end of the chapter.

$$a * C_A + b * C_B \Rightarrow c * C_C \tag{4.1}$$

$$r_T = k_o * e^{-E/(R*T)} \tag{4.2}$$

$$r_x = r_T * [C_A]^{\alpha} * [C_B]^{\beta} \tag{4.3}$$

$$SP_B = SP_A * R_{B/A} \tag{4.4a}$$

$$R_{B/A} = \frac{b * MW_B}{a * MW_A} \tag{4.4b}$$

The mass flow ratio for chemical reactants is the ratio of the molecular weights multiplied by the respective stoichiometric coefficient for each reactant.

Not shown in the equations is the crucial role temperature often plays in determining reaction selectivity. High temperatures can trigger reverse reactions (where product is converted back to reactant), product degradation (where product quality suffers), and side reactions (where reactants are converted into waste products that require utilities and processing time to separate and treat). Besides production rate, reactor temperature often dictates product quality and process efficiency besides production rate. Reactor temperature loops require greater measurement accuracy and tighter control than most other process loops. Fortunately the process dynamics of a well-mixed reactor facilitate a controller gain much higher than realized by practitioners. The PID gain can be 50 or more because the gain is proportional to a large primary time constant to dead time ratio or the inverse of the product of a small open loop integrating process gain and dead time as detailed in Section 1.3.

4-4. Biological Reactions

The operating condition requirements for optimum cell health, cell growth rate, and product formation rate increase as the cell sophistication increases. Yeast cells have the simplest requirements. For many fermenters used in fuel and beverage production, only the temperature is controlled. In fungal and bacterial bioreactors, dissolved oxygen, pH, and temperature are controlled and a glucose addition is scheduled.

In mammalian cell bioreactors a dielectric spectroscopy probe is used as an online analyzer to measure viable cell concentration. Besides glucose, glutamine addition is scheduled. An optimum ratio of glutamine to glucose exists that depends upon batch conditions and time. Most bioreactors currently do not try to reach and maintain this optimum online. Flexible at-line analyzers using electrochemistry, digital imaging, freezing point depression, and photometry (e.g., Nova BioProfile Flex) can measure in 2-8 minutes the cell density, cell diameter, cell viability, osmolality, substrates (glucose and glutamine), and many byproducts such as carbon dioxide, lactate, and ammonia that inhibit cell growth and product formation. These analyzers are used in bench-top and pilot plant bioreactors but have not currently been installed as at-line analyzers in many production lines. Near infrared (NIR) spectroscopy can be used to provide online measurements of substrate, byproduct, and product concentrations but these analyzers generally require more sophisticated calibration knowledge and practices. Mass spectrometers can measure the amount of oxygen and carbon dioxide in the bioreactor off-gas for oxygen uptake rate and carbon dioxide production rate calculations. Mass spectrometers are routinely used and are extremely accurate if properly installed and maintained.

Bio-reaction kinetics in the literature has a common form that uses a series of multiplicative factors each with a fractional value (i.e., value between 0 and 1.0). Each factor models the effect of an operating condition on viable cell growth rate and product or byproduct formation rate. The multiplicative form is consistent with the multiplicative effects in cell responses, especially as the complexity increases from yeast to fungal to bacterial to mammalian.

This large number of multiplicative factors is not addressed in the literature for parameter identification. The methods published to date on automated parameter estimation focus on equations that are a summation of terms rather than a multiplication of terms. Most search techniques cannot handle the nonlinear interactive effects of multiplicative factors. A log transformation could enable an additive form of terms for parameter estimation but practical demonstrations of this capability has not been achieved probably due to the large number of factors, the high sensitivity to operating conditions, and the sparse number of batches due to batch cycles times of days to weeks.

The equations for mammalian cells have the most factors. As the number of factors increases, so does the challenge to keep the individual factors closer to 1.0 because of the multiplicative effect. Most bioreactors are not operating at the optimum conditions that maximize these factors. Lab optimized DCS systems (e.g., Broadley-James BioNet) with advanced control and data analytics integrated with advanced bioreactor modeling objects (e.g., MYNAH MiMiC) offer the ability for the biochemist to find and control these factors at the optimums by virtual experimentation that can lead to better focused and valuable laboratory runs.

The multiplicative form of kinetic terms indicative of cell response poses nonlinear interactive effects, making difficult the automated parameter identification of all of terms for complex cells (e.g., 7 or more terms for mammalian cells).

The growth rate of viable cells (r_v) (g/L per hr) is the specific growth rate (μ_v) (1/hr) multiplied by the viable cell concentration (X_v) (g/L) (Equation 4.5). The specific growth rate depends upon the maximum specific growth rate (μ_{vmax}) multiplied by individual rate factors (r_{vi}) (Equation 4.6). These kinetic models are termed extracellular because they predict the measurable effects and consequences of operating conditions outside the cell. Kinetic models for predicting what occurs inside the cell are much more subjective and have an inherent difficulty of not being able to measure the effects and consequences of operating conditions inside the cell. Models can only be developed for inputs and outputs that can be measured; otherwise they do not meet the scientific method criterion of validation of the equations by measurements. The nomenclature with units is defined at the end of the chapter.

$$r_v = \mu_v * X_v \tag{4.5}$$

$$\mu_v = \mu_{v\max} * r_{vs_1} * r_{vs_2} * r_{va} * r_{vb} * r_{vO2} * r_{vH^+} * r_{vT} \tag{4.6}$$

There are similar expressions for product (j) or byproduct (j) (e.g., lactate and ammonia) formation rate for operating condition (i). These can be generalized to the following format (Equations 4.7a and 4.7b). The number of factors can be small to start with and can be increased as knowledge is gained.

The kinetics equation form used for cell growth rate can be used for product formation rate.

$$\mu_j = \mu_{j\max} * \prod r_{ji} \tag{4.7a}$$

$$r_j = \mu_j * X_v \tag{4.7b}$$

The individual dimensionless multiplicative factors (r_{ji}) for the effect of bioreactor concentrations (X_i) (e.g., glucose, glutamine, ammonia, and lactate) can be approximated by a factored Michaelis-Menten equation with an inhibition parameter (K_{1ji}) and a limitation parameter (K_{2ji}) (Equation 4.8). The parameters are in the same units as the concentrations (e.g., g/L). For low concentrations, the inhibition term approaches 1.0 and the limitation term is proportional to the ratio of concentration to the limitation parameter. The limitation term is the well-known Monod equation in biochemistry. For high concentrations, the limitation term approaches 1.0 and the inhibition term is proportional to the ratio of the inhibition parameter to the concentration. If there is no limitation effect, the limitation parameter is simply set equal to zero (K_{2ji} = 0 g/L). If there is no inhibition effect, the inhibition parameter is set to be at least two orders of magnitude larger than the maximum corresponding concentration (e.g., K_{1ji} = 100 g/L). The nomenclature with units is defined at the end of the chapter.

The application of gain factors and biases enables a more flexible setup for modeling in virtual plants. A filter can be added to the individual rate factors to simulate the lag in cell response. The expressions used for the rate factors vary considerably with cell and product type. The Michaelis-Menten equations have been shown to provide a generic solution for modeling the effect of concentration and even offer a notable ability to model the results of complex intracellular kinetics. The exception is the Nielsen-Villadsen equation that is the Michaelis-Menten equation applied to hydrogen ion concentration to model the effect of pH. The extreme sensitivity of pH to hydrogen ion concentration makes parameter fitting impractical.

$$r_{ji} = \left[\frac{K_{1ji}}{X_i + K_{1ji}}\right] * \left[\frac{X_i}{X_i + K_{2ji}}\right] \quad (4.8)$$

The Michaelis-Menten equation provides the generic terms for the limitation and inhibition effects of concentration (except for hydrogen ions) on cell growth rate or product formation rate.

While the literature has shown the use of the Nielsen-Villadsen equation to model the effect of pH (Equation 4.8 with hydrogen ion concentration used instead of cell concentration) and the Arrhenius Equation 4.2 to model the effect of temperature, the parameters are too sensitive requiring a precision to six or more decimal places. A clue to this problem was gained when the equation given for the effect of pH and temperature in an example gave bizarre results simply because the book had rounded off the parameter to four decimal places. Alternate equations have been developed with parameters that are more intuitive and easier to initialize besides requiring much less precision.

The Convenient Cardinal Equations 4.9a and 4.9b for pH and temperature, developed in the February 1995 paper "Convenient Model to Describe the Combined Effects of Temperature and pH on Microbial Growth" published in *Applied and Environmental Microbiology*, can give the same curve on a plot of a kinetic rate factor versus pH or temperature as the Nielsen-Villadsen equation for pH and the Arrhenius equation for temperature. Equations 4.9a and 4.9b offer the important advantage of using parameters that are readily recognizable and easily set with a minimal precision requirement. Figures 4.1 and 4.2 show examples of such curves obtained from these equations. The parameters provide reasonable results for a precision of just one decimal place. The parameters are minimum, maximum, and optimum pH or temperature. The nomenclature with units is defined at the end of the chapter.

$$r_{H^+} = \left[\frac{(pH - pH_{min})*(pH - pH_{max})}{(pH - pH_{min})*(pH - pH_{max}) - (pH - pH_{opt})^2}\right] \quad (4.9a)$$

$$r_T = \left[\frac{(T - T_{max})*(T - T_{min})^2}{(T_{opt} - T_{min})*[(T_{opt} - T_{min})*(T - T_{opt}) - (T_{opt} - T_{max})*(T_{opt} + T_{min} - 2*T)]}\right] \quad (4.9b)$$

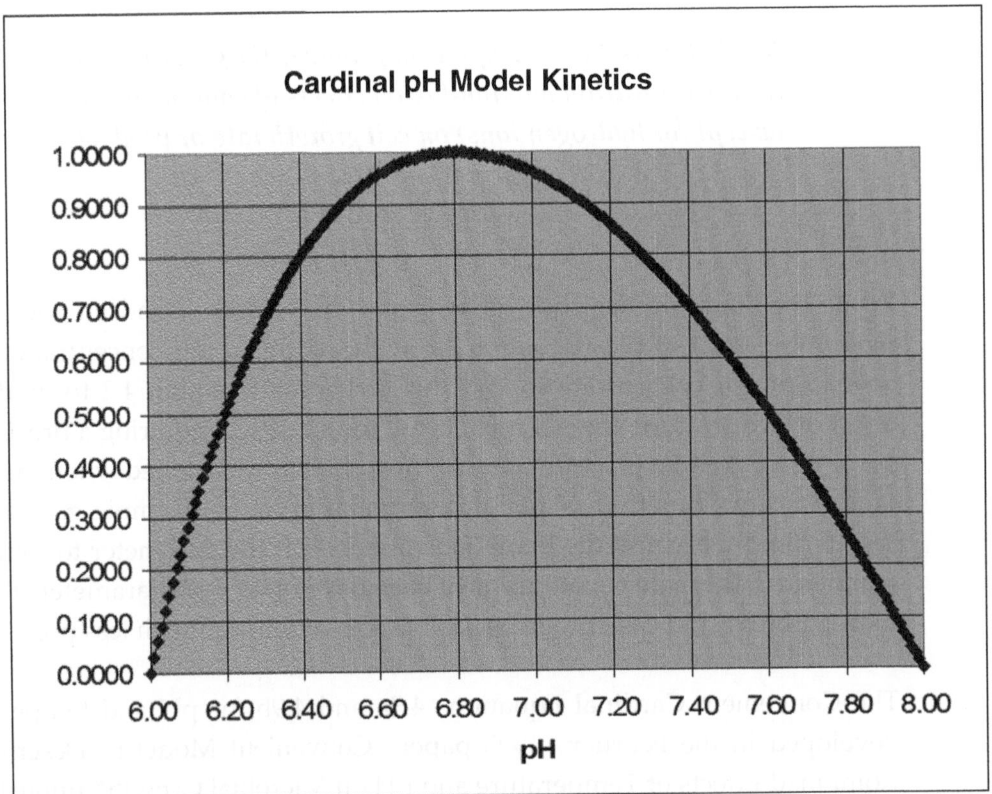

Figure 4.1a – Cell growth rate versus pH obtained by Convenient Cardinal Equation

Mammalian cells are the most sensitive to pH and temperature. Maintaining the maximum cell growth rate or product formation rate may require control within a few hundredths of a pH and a few hundredths of degree Centigrade. The moderating effect of carbon dioxide and the weak acids and bases, coupled with the extremely slow formation of acid and base byproducts, enable much tighter pH control in bioreactors than is realizable in chemical reactors. The extremely slow changes in metabolic heat enable much tighter temperature control. Changes in reagent demand or cooling are so slow that tuning for

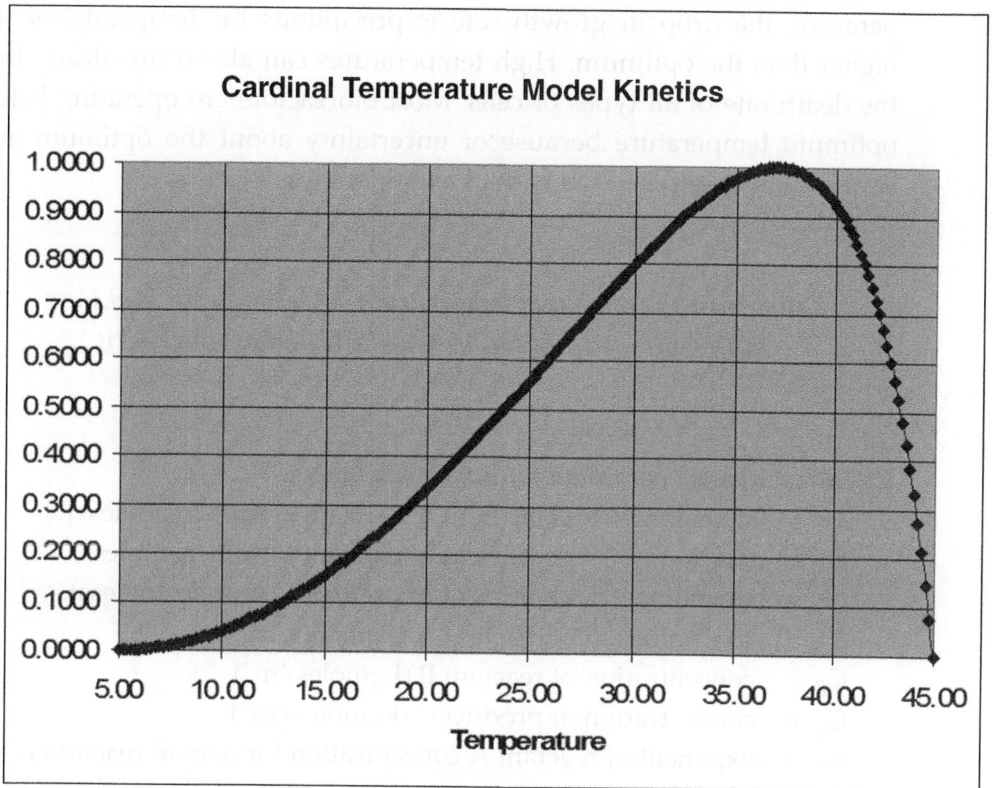

Figure 4.1b – Cell Growth Rate versus Temperature obtained by Convenient Cardinal Equation

> *The Convenient Cardinal Equations provide equations for readily visualized and initialized parameters for the effect of pH and temperature on cell growth rate or product formation rate.*

disturbance rejection is not important. The setpoint response, however, is critical. Setpoint changes occur at the start of the batch and for shifts in pH and temperature to promote cell growth or product formation. The overshoot of setpoint must be less than a tenth of a pH or degree Centigrade. The setpoints for optimum cell growth and for product formation are different. The emphasis on cell growth and product formation typically occurs early and late in the batch, respectively.

The approach to the optimum temperature must be cautious because the downside is extreme. While the increase in growth rate is gradual with temperature, the drop in growth rate is precipitous for temperatures slightly higher than the optimum. High temperatures can also dramatically increase the death rate of all types of cells. Most bioreactors are operating below the optimum temperature because of uncertainty about the optimum and the severe consequences of too high of a temperature.

> ***Going past the optimum temperature for cell growth is akin to falling off a cliff in terms of cell life and productivity.***

Chemical kinetic parameters for two reactants:

a = stoichiometric coefficient of reactant A (dimensionless)
b = stoichiometric coefficient of reactant B (dimensionless)
c = stoichiometric coefficient of product C (dimensionless)
C_A = concentration of reactant A (kgmoles/m^3)
C_B = concentration of reactant B (kgmoles/m^3)
C_C = concentration of product C (kgmoles/m^3)
α = exponent of reactant A concentration for overall reaction rate (dimensionless)
β = exponent of reactant B concentration for overall reaction rate (dimensionless)
MW_A = molecular weight of reactant A (dimensionless)
MW_B = molecular weight of reactant B (dimensionless)
$R_{B/A}$ = mass flow ratio reactant B to reactant A (dimensionless)
SP_A = mass flow setpoint of reactant A (kgmoles/hr)
SP_B = mass flow setpoint of reactant B (kgmoles/hr)
k_o = Arrhenius equation pre-exponential factor (kgmoles product/hr/(α*β kgmoles reactants))
r_T = Arrhenius equation reaction rate (kgmoles product/hr/(α*β kgmoles reactants))
E = Arrhenius equation activation energy (kJ/kgmole)
R = Arrhenius equation universal gas law coefficient (kJ/kgmole/^{o}K)
T = Arrhenius equation reaction temperature (^{o}K)
r_x = overall reaction rate (kgmoles product/hr)

Biological kinetic parameters for cell growth rate or product formation rate:

r_{ji} = cell growth rate or product formation rate factor due to component i (g/L per hour)
μ_j = specific cell growth rate or product formation rate (g /g cell per hr)
$\mu_{j\max}$ = maximum specific cell growth rate or product formation rate (g /g cell per hr)
X_v = viable cell concentration (g cells/L)
X_i = component i concentration (g /L)
K_{1ji} = inhibition parameter due to component i (g/L)
K_{2ji} = limitation parameter due to component i (g/L)
$pH_{\max}$ = maximum pH for cell growth or product formation (pH)
$pH_{\min}$ = minimum pH for cell growth or product formation (pH)
pH_{opt} = optimum pH for cell growth or product formation (pH)
$T_{\max}$ = maximum temperature for cell growth or product formation (^oC)
$T_{\min}$ = minimum temperature for cell growth or product formation (^oC)
T_{opt} = optimum temperature for cell growth or product formation (^oC)

5
Reactor Control Systems

5
Reactor Control Systems

5-1. Single Phase Chemical Reactors

For reactions that are all in one phase (gas, liquid, or solid), inventory control (pressure or level control) cannot automatically adjust a reactant flow for changes in conversion (amount of product or byproduct formed per unit of reactant added) to prevent an excess or deficiency of reactant in a product phase (e.g., liquid) that is the opposite of the reactant phase (e.g., gas). The exact ratio of reactants needed is determined by the stoichiometric ratio as exemplified by Equation 4.4a in Section 4.3 for a simple reaction.

For single phase reactors, the use of analyzers is more important to control reactant stoichiometry for maximizing yield beyond what tight temperature control can do. Section 5.2 will describe the inherent excess reactant control possible in multiple phase reactors that helps reduce the dependence on concentration control.

> ***Single phase chemical reactors have a greater dependence on analyzers to maximize yield since there is not the inherent prevention of excess reactants by pressure or level control.***

For well mixed single phase reactors the largest sources of improper reactant concentration leading to excess reactant are errors in reactant flow measurement and changes in reactant composition. Note that a deficiency in one reactant concentration creates an excess of another reactant. Coriolis meters can be used to provide the greatest mass flow measurement precision and rangeability with density correction for any changes in reactant feed concentration. Consequently, Coriolis meters on reactant feeds eliminate most of the sources of reactant unbalances if the mass flow ratios are correct and coordinated to maintain reaction stoichiometry.

Coriolis meters in reactant feeds are the best way to ensure correct stoichiometric ratio if the feed setpoints are changed in unison with the correct ratio and the same closed loop response.

If the density of the excess reactant is significantly different than the density of the other components in the reactor, a Coriolis meter in the recirculation line can provide an inline inferential measurement of the excess reactant concentration. Inline composition measurements by means of sensors in a vessel or pipeline provide a measurement in a few seconds whereas at-line analyzers with sample systems can have 30 or more minutes of dead time due to sample delay and analyzer cycle times. If inline or at-line analyzers are not feasible, lab analysis results must be communicated as quickly as possible to the control system with the time that the sample was taken. The composition measurement of an excess reactant should be used to provide a correction of the ratio of reactants. Coriolis meters are most accurate on liquid streams but can used on gas streams of sufficiently high density due to pressure or molecular weight.

If changes in excess reactant concentration cause density changes greater than 1% and are at least 100 times larger than other density changes, a Coriolis meter in a recirculation line can be used as an inline analyzer for excess reactant concentration.

For continuous reactors with liquid reactants and liquid product, a level loop controls the inventory and thus the time available for reaction by manipulating the discharge product flow (Figure 5.1a). If the desired reactor residence time multiplied by the total reactant flow is scaled and used as the setpoint of the level controller, the level loop helps maintain a constant residence time. The reactor discharge flow is not used in the residence time calculation, to avoid the creation of a second loop within the level loop where a change in level PID output would cause a change in level PID setpoint which would cause a change in level PID output and so forth. The reactant feed setpoints are used instead of reactant flow measurements for the same reasons that setpoints are used in feedforward flow signals.

Batch reactors don't have a level loop and the concept of a residence time is replaced with a batch time for available time for reaction. The rising level of a batch reduces any self-regulation that might exist; changes in heat or mass transfer with temperature and concentration affect the open loop responses of temperature and product composition necessitating the use of higher controller gains to provide sufficient closed loop regulation. Integral action is also more likely to produce overshoot of setpoint.

A liquid level controller with a setpoint proportional to production rate can help provide the necessary residence time for liquid reactants in a continuous reactor.

Temperature loops control the energy balance and the reaction rate through the Arrhenius equation detailed in Chapter 3. A cascade temperature control system offers the greatest linearity and responsiveness to coolant pressure and temperature upsets (Figure 5.1a). The reactor temperature PID manipulates the setpoint of a jacket inlet temperature control that in turn manipulates makeup coolant flow. The coolant exit flow, such as cooling tower water (CTW) return flow, equals the coolant makeup flow to the jacket by piping design and in some case by pressure control in the recirculation system. The resulting constant jacket coolant flow eliminates the increase dead time, process gain, and fouling from a decrease in jacket flow. An enhanced PID is used for the jacket inlet temperature loop to prevent limit cycling from valve stick-slip. The relative merits of various temperature system designs are discussed in Chapter 6 on cooling and heating systems.

Temperature control determines the reaction rate via the Arrhenius equation.

Cascade control of reactor to jacket temperature isolates the reactor temperature PID from most of the nonlinearities and disturbances associated the jacket temperature response.

For batch reactors, as previously mentioned, there is no level control and hence no residence time control but otherwise the temperature and gas pressure control systems shown in this chapter for continuous liquid reactors are generally applicable. As per Chapter 1 the same tuning rules can be used since a large primary process time constant enables continuous liquid reactor composition and temperature loops to be treated as having a near-integrating response that is similar to the integrating response of batch reactors.

The temperature and gas pressure control systems and PID tuning rules for batch reactors and liquid continuous reactors are similar.

For pure batch operations where flows are sequenced and scheduled based on batch times and totals the concentration endpoint is achieved by accurate charging of reactants and by a batch cycle time long enough to insure complete reaction. The simultaneous feeding of reactants by flow loops whose setpoints can be manipulated, termed fed-batch control or semi-continuous control, opens up opportunities for the use of concentration control as the batch progresses. The control schemes for concentration control of excess reactants in continuous liquid reactors, by correction of the reactant ratio shown in this chapter can be extended to fed-batch reactors. However, for non-reversible reactions the concentration control of product, as the batch progresses requires the translation of the controlled variable because the product concentration response is only in one direction. A translation to a rate of change of product concentration; slope of the batch product concentration profile provides a controlled variable that can increase and decrease based on feedback action. The computation provides an indication of product formation rate and yield and enables a prediction of batch endpoint. See Section 5.8 on fed-batch profile control for a discussion of the opportunities for more repeatable and predictable batch composition profiles and a faster batch cycle time.

Analyzers and inferential measurements can provide composition control to correct the ratio of reactants (Figure 5.1a). The leader reactant flow controller setpoint is multiplied by a ratio factor and added to the composition controller output. The setpoint is used instead of the actual flow measurement to reduce noise from the process and sensor (e.g., pressure and velocity profile fluctuations in differential head meters) and cycling from valve backlash or

Reactant concentration control systems for continuous liquid reactors can be extended to fed-batch reactors. However, product concentration control of non-reversible reactions requires a translation of the controlled variable to be the slope of the product concentration profile.

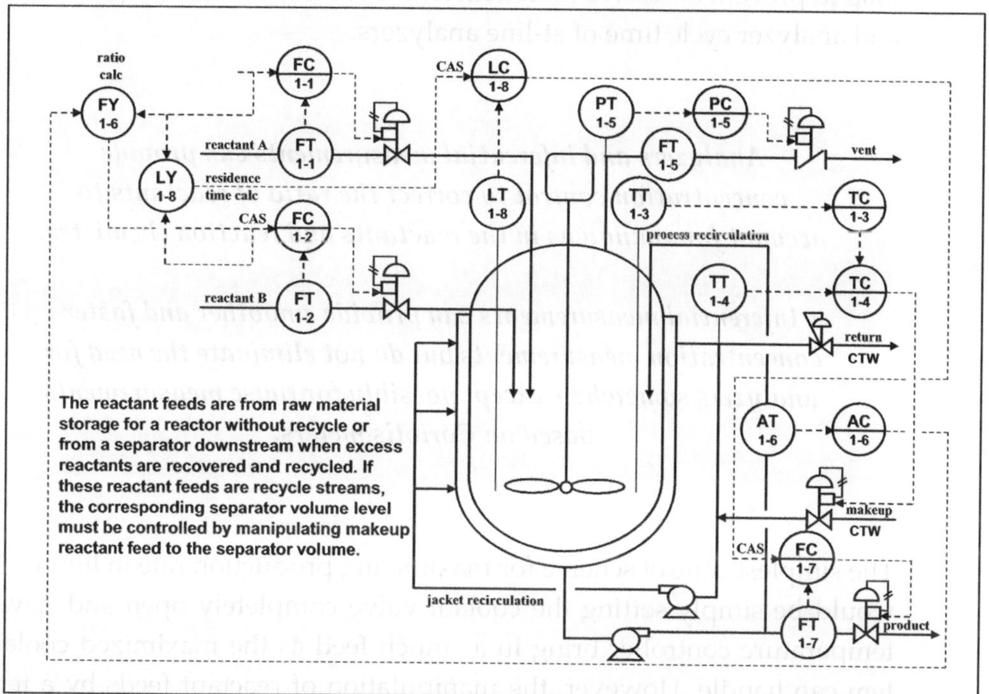

Figure 5.1a – For a liquid reactor, level control sets reaction time via residence time, temperature control sets reaction rate via energy, and composition control enforces the stoichiometric ratio

stick-slip and to enable better timing of the flow adjustments for production rate changes. However, the valve must be able to reach the flow setpoint by proper valve sizing and a valve design that prevents the valve from getting stuck or plugged. The reactant flow controllers are tuned for the same closed loop time constant to eliminate transient unbalances in the stoichiometric ratio of reactants added. A feedforward summer is used because both the process gain and the process time constant have an inverse relationship to reactant flow as discussed in Chapter 1 and detailed in Appendix F. Since the maximum controller gain is proportional to the ratio of process time constant to process gain, the effect of flow cancels out in the tuning of the controller.

At-line analyzers can be multiplexed with automated sample systems so one analyzer can service several reactors to help justify the cost. Inferential measurements can make the concentration measurement faster and smoother and less susceptible to missing or spurious results and noise. Inferential measurements except possibly for those from Coriolis meters must be periodically corrected by analyzer measurements ideally in the process but possibly in the lab. An enhanced PID (described in Appendix E) is used in the composition loop to prevent excessive reset action and cycling from the large sample delay and analyzer cycle time of at-line analyzers.

> ***Analyzers and inferential measurements can provide concentration control to correct the ratio of reactants to account for unknowns in the reactants and reaction chemistry.***
>
> ***Inferential measurements can provide smoother and faster concentration measurements but do not eliminate the need for analyzers somewhere except possibly for those measurements based on Coriolis meters.***

The simplest control scheme for maximizing production rate in liquid reactors would be simply setting the coolant valve completely open and having the temperature controller bring in as much feed as the maximized coolant system can handle. However, the manipulation of reactant feeds by a temperature controller for maximization of production rate in liquid reactors causes inverse response for feeds colder than the reaction temperature and introduces a large lag that is the composition time constant of the vessel volume. For fast gas reactions, the residence time and consequently the composition time constant are small enough for this scheme to be effectively used, particularly when reactions are highly exothermic with a large heat generation that overwhelms any cooling effect of feeds.

> ***The manipulation of reactant feeds for temperature control in liquid reactors does not work due to excessive inverse response and lag in the liquid composition response. The control scheme is productive for fast gas reactions.***

In a first order plus dead time approximation for the temperature response, all time constants smaller than largest time constant are converted to equivalent dead time. The thermal process time constant for temperature control should be the largest time constant and the primary process time constant in the loop. The direct manipulation of feed rate for temperature control coupled with a slow reaction rate can cause a composition response time constant larger than the thermal time constant. The result is that a fraction of the thermal time constant becomes effectively dead time, causing a decrease in the maximum allowable PID gain and an increase in the minimum reset time and the peak and integrating errors for load disturbances.

A composition response in series with the thermal response in liquid reactors creates excessive dead time and deterioration in loop performance.

In a liquid reactor a coated temperature sensor or a sensor tip that does not extend past the nozzle into the vessel contents or a sensor hidden behind a baffle, could cause a measurement time constant larger than the thermal time constant. A sensor in a ceramic protection tube in a gas reaction phase or a sensor in a baffle with a glass coating in a small liquid volume will also have an excessive measurement lag due to the poor surface thermal conductivity of the sensor installation.

The slow composition response from manipulation of liquid reactant feeds causes slow correction and the slow sensor causes slow recognition of a disturbance. In either case, thermal time constants become effectively dead time and control severely deteriorates.

A sensor that is coated or is installed in a stagnant zone (e.g., behind a baffle) will add an excessive measurement lag.

The temperature sensor location is also important for the secondary jacket temperature loop. The sensor should be located in the coolant recirculation line rather than in the jacket. The higher velocities and turbulence in the pipe-

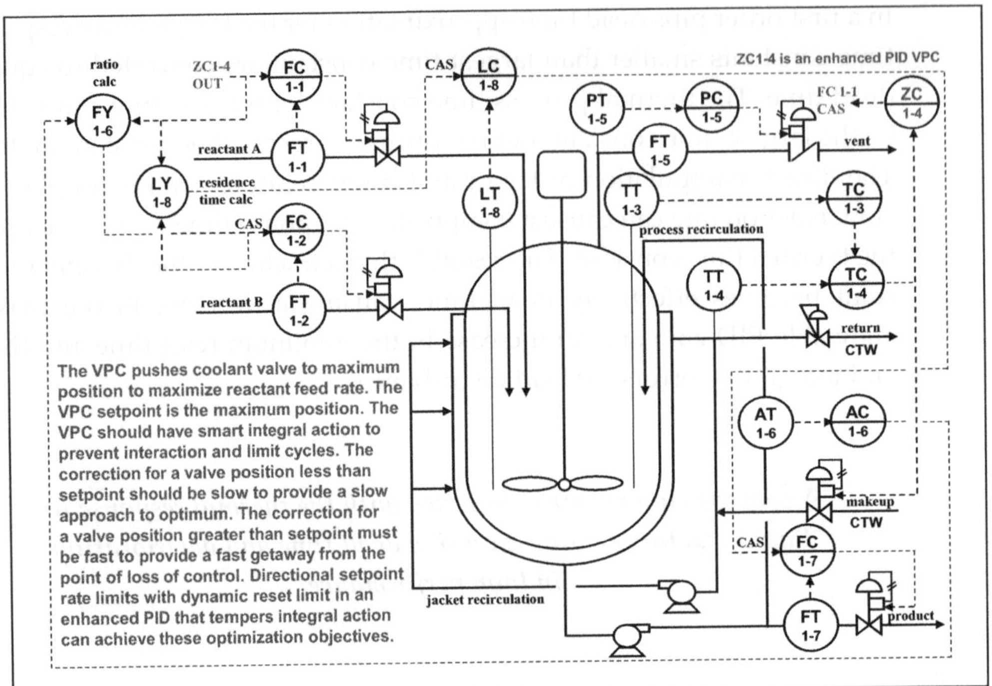

Figure 5.1b – For a liquid reactor, the production rate can be maximized by a VPC (ZC1-4) that increases reactant feed till the jacket temperature valve reaches maximum position

line provide a faster measurement with fewer fluctuations from level and phase changes and cold or hot spots from product sticking on the reactor wall.

Production rate can be maximized by the use of a valve position controller (VPC) monitoring coolant valve position (Figure 5.1b). The VPC setpoint is the maximum desirable valve position, and the VPC process variable is the jacket temperature controller output. The use of actual valve position is unnecessary if the coolant valve has a digital positioner. The maximum throttle valve position setpoint keeps the coolant valve near a point on the installed flow characteristic that has sufficient slope (valve gain) to correct for disturbances. The output of the VPC trims the setpoint of the "leader" reactant flow controller. An enhanced PID with external reset feedback (dynamic reset limiting) for the VPC eliminates limit cycles from coolant valve deadband, reduces interaction between the VPC and the jacket temperature controller, and enables smoother optimization with faster correction for large disturbances by directional move suppression. Directional move suppression by means of slow up and fast down rate limits on the manipulated feed flow setpoints enables a gradual optimization and fast correction for abnormal conditions.

A reactant feed flow setpoint rate limit should be fast for correcting an increase in coolant valve position to prevent "running out of valve" for high heat releases. The setpoint rate limit should be slower for the opposite direction to provide a more gradual optimization. External reset feedback in the PID automatically prevents integral action from driving the VPC PID output faster than rate limits allow or than the feed flow can respond. To see the increase in production rate in fed-batch operation from a higher feed rate, either the batch cycle time must be allowed to decrease or the batch mass allowed to increase.

> ***A VPC, whose setpoint is the maximum coolant throttle valve position, can maximize feed rate smoothly with less interaction and better response to disturbances by the use of an enhanced PID and reactant feed setpoint rate limits to provide directional move suppression.***

The reactant feeds shown in Figures 5.1a and 5.1b could be fresh reactant streams from raw material storage or recycle streams when excess reactants are recovered in downstream separators. For the recycle case, the level in the corresponding downstream separator volume must be controlled by the manipulation of makeup reactant flow. For an excess reactant that is a heavier or lighter component (has a lower or higher boiling point) than the product, the separator volume that is the source of the reactant recycle flow, is a column sump or distillate receiver, respectively.

There must be a flow controller somewhere in the path of a recycle stream from reactor discharge to reactor feed to prevent the "snowballing effect", an escalating divergence of excess reactant concentration from any upset. The flow controllers on the reactant feeds in Figures 5.1a and 5.1b serve this purpose when these are recycle streams.

> ***When excess reactants are recovered and recycled to the reactor from downstream separators, a flow controller must exist in each recycle path to prevent the "snowballing effect".***

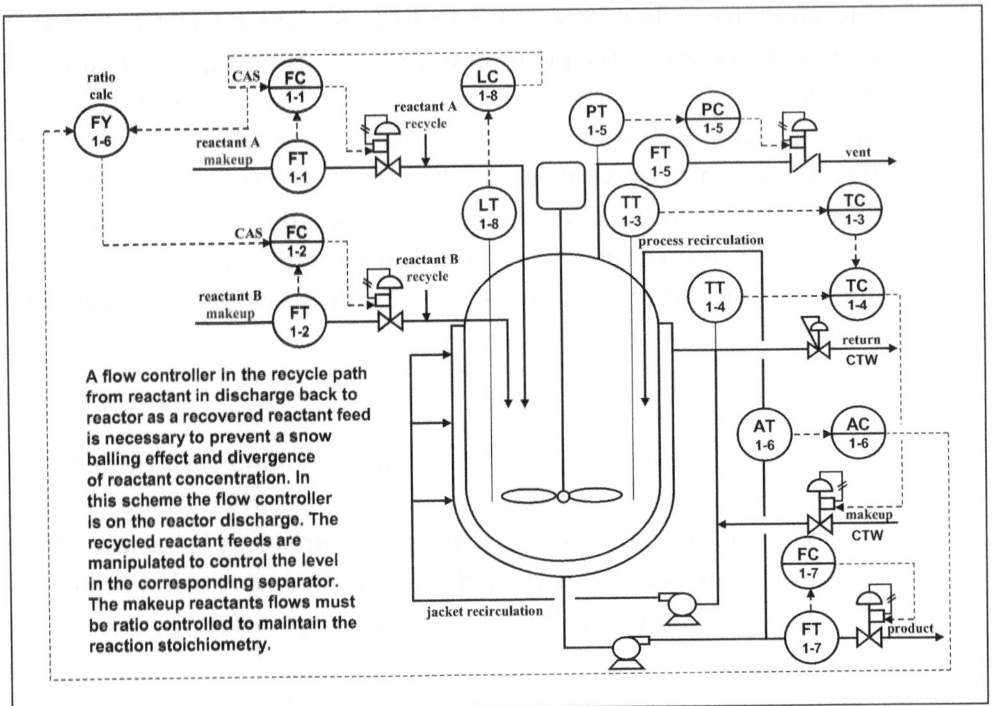

Figure 5.1c – For a liquid reactor with recycle of recovered reactants set by downstream separator level controllers, the makeup reactant flows must be ratioed to maintain stoichiometry

An alternate control scheme for recycle systems is shown in Figures 5.1c and 5.1d where the reactor level controller manipulates a makeup reactant flow. The reactor discharge flow controller sets production rate and prevents the "snowballing effect." The makeup reactant feeds must be ratio controlled to maintain stoichiometry. If reactant B flow is left on local flow control, the reactor concentration of reactant B will go low for an increase in production rate or a disturbance that caused a decrease in reactor level. An accurate analyzer is more important in this control scheme to correct the ratio because the lack of inherent self-regulation can cause a continual error in the stoichiometry.

If the makeup reactant flows are set by the reactor control system rather than corresponding downstream separator volume level controller, the makeup reactant flows must be accurately ratio controlled to maintain stoichiometric ratio for changes in reactor level.

Figure 5.1d shows a reactor with recycled and makeup reactant combined as feed streams that can use a valve position controller to push the reactor coolant valve to the maximum throttle position to gradually increase the reactor discharge flow and hence production rate. The optimization must be extremely slow to prevent the "snowballing effect." The use of directional move suppression that offers a smooth gradual approach to an optimum and a fast recovery from an upset is extremely important here to prevent a problem.

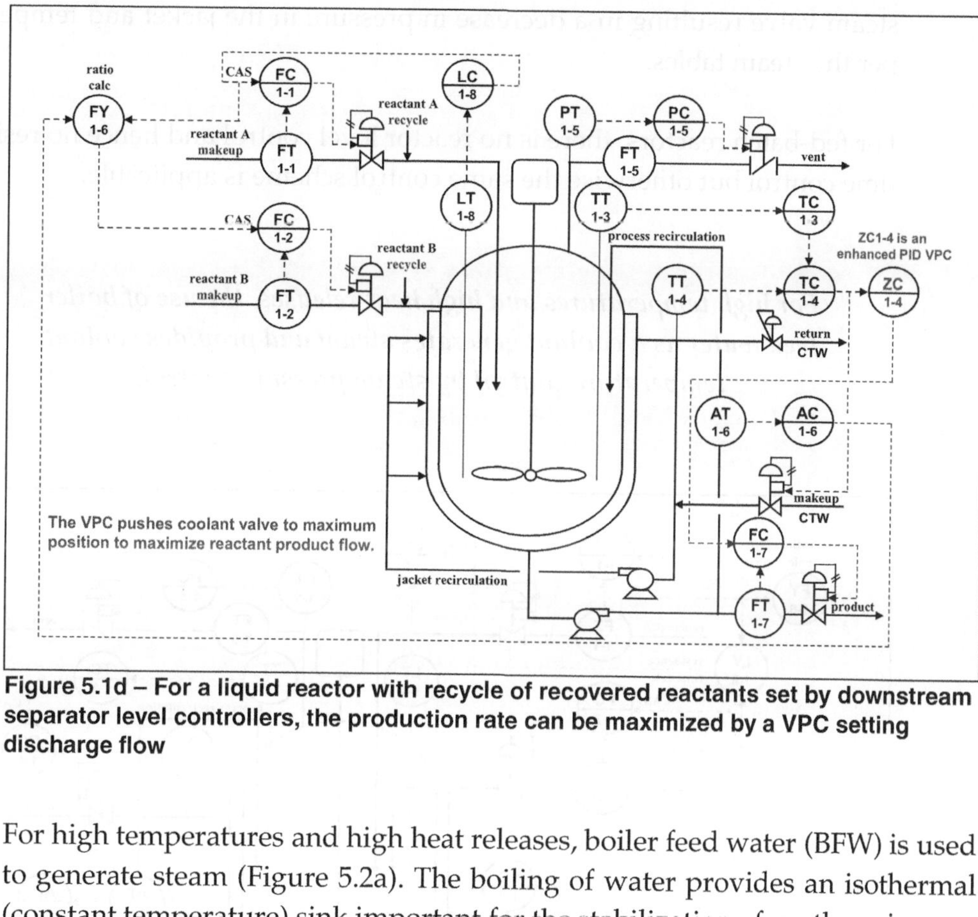

Figure 5.1d – For a liquid reactor with recycle of recovered reactants set by downstream separator level controllers, the production rate can be maximized by a VPC setting discharge flow

For high temperatures and high heat releases, boiler feed water (BFW) is used to generate steam (Figure 5.2a). The boiling of water provides an isothermal (constant temperature) sink important for the stabilization of exothermic reactors. The jacket pressure and level controller maintain inventory of the vapor and liquid phases. An increase in heat release causes an increase in steam generation. The pressure controller lets out the additional steam keeping the jacket pressure and jacket temperature constant. Thus, an increase in heat release for a constant reactor temperature does not increase jacket temperature as would be the case for liquid coolant systems.

If BFW was not used for high heat releases, the coolant temperature would approach the reactor temperature reducing the temperature difference across the heat transfer surface that is the driving force for heat transfer. The generation of steam keeps the temperature difference constant if the reactor temperature has not changed. If the temperature of the reactor changes the temperature controller changes the setpoint of the pressure controller, which changes the steam temperature. An increase in reactor temperature decreases the steam pressure setpoint. The pressure control in turn opens the outlet steam valve resulting in a decrease in pressure in the jacket and temperature per the steam tables.

For fed-batch reactors, there is no reactor level control and hence no residence time control but otherwise the same control scheme is applicable.

For high temperatures and high heat releases, the use of boiler feed water as a coolant generates steam and provides coolant temperature control by steam pressure control.

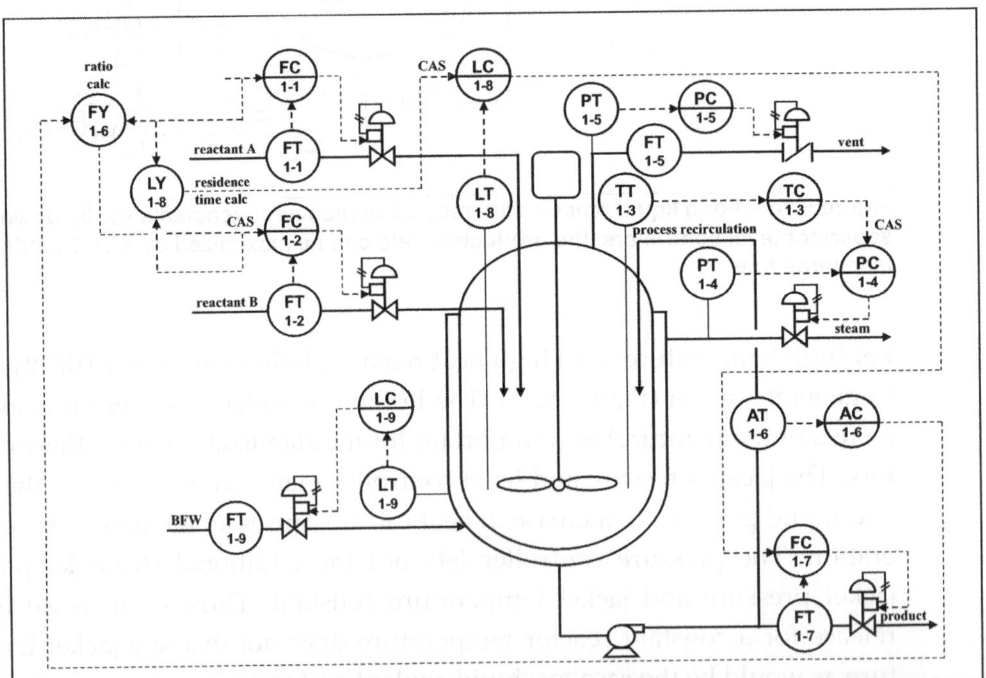

Figure 5.2a – For a high temperature liquid reactor, coolant in Figure 1a is replaced with the boiling of water to provide a constant temperature heat sink that helps stabilize highly exothermic reactions

Production rate can be maximized by the use of a valve position controller (VPC) monitoring BFW and steam valve positions (Figure 5.2b). The VPC setpoint is the maximum desirable valve position, and the VPC process variable is the highest of the level and pressure controller outputs. Differences in the nonlinearity of the coolant and vent valve can be eliminated by signal characterization so the valve curve slope (gain) and setpoint are about same for either valve. The output of the VPC trims the setpoint of the "leader" reactant flow controller. An adaptive PID for the VPC could switch tuning settings based on which valve is being controlled. An enhanced PID with dynamic reset limiting for the VPC eliminates limit cycles, reduces interaction between the VPC and the level and pressure controllers, and enables smoother optimization with faster correction for large disturbances.

The production rate of BFW coolant systems can be maximized by a VPC that pushes the BFW level control valve and steam pressure control valve to the maximum throttle valve position.

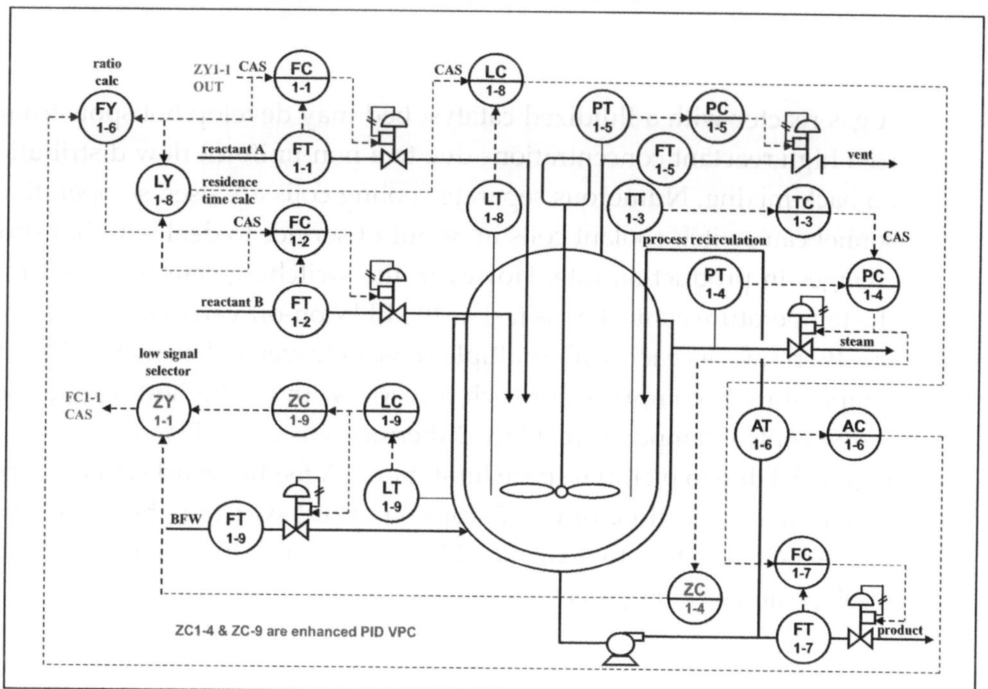

Figure 5.2b – For a high temperature liquid reactor, the production rate can be maximized by a VPC that increases reactant feed till the BFW or steam valve reaches maximum position

For gas reactants and a gas product, a pressure loop controls the material balance and provides the time available for reaction at a given production rate by manipulating the discharge product flow to balance the total feed flow (Figure 5.3). The residence time for these fast reactions is small (e.g., a few seconds) but still must be kept above a low limit. A fluidized catalyst bed is used to promote reaction rate. If a temperature loop manipulates the leader gas reactant flow, the production rate is automatically maximized by the temperature and pressure controllers for a given cooling rate established by BFW flow and the number of coils in service as discussed below. Valve position control is not needed unless a control strategy is added to manipulate the BFW. Direct manipulation of feed rate by the temperature control is possible in gas reactors because the additional time lag for composition response is negligible due to the small residence time and the inverse response at the control points is negligible due to the fast reaction, high heat release, and catalyst heat capacity.

Temperature control of fluidized bed gas reactor by manipulation of feed rates will inherently maximize feed rate for a given cooling capability with no appreciable inverse response or composition response lag.

A gas reactor with a fluidized catalyst bed may develop hot spots from localized high reactant concentrations due to a non-uniform flow distribution and no back mixing. Numerous separate cooling coils are used so operations personnel can switch coolant coils in or out of service to deal with hot spots and changes in production rate. However, the switching causes a disturbance to the temperature controller as fast as the BFW on-off valves can move. Numerous thermowells each with multiple sensors traverse the reactor. The average temperature is computed for each traverse with the highest average selected as the control temperature. Only 3 thermowells and BFW coils are shown in Figure 5.3 due to pictorial space limitations. A feedforward signal can provide preemptive correction for the disruption of coil switching by means of a gain and velocity limit set to match the BFW on-off valve installed flow characteristic slope and stroking time.

Fluidized catalyst bed reactors use the highest of the average bed temperature at various distances in the flow direction to control hot spots.

A feedforward signal based on the installed characteristic and stroking time of the BFW on-off valves can be used to reduce the temperature upset from the switching of cooling coils.

The plug flow of reactants through the reactor provides a tight residence time distribution. To provide a residence time greater than the reaction time for the greatest production rate, the gas volume must be large enough and the flow distribution uniform enough for the reaction to go to completion.

Changes in discharge composition are mainly due to errors in the flow measurement, hot spots triggering side reactions, or insufficient radial mixing. An excess of one reactant is often used to ensure complete conversion of other reactants. An at-line analyzer on the gas product can be used to correct the gas reactant ratio to improve yield by reducing excess reactant.

The leader reactant flow multiplied by a ratio factor is the feedforward signal for the composition controller. A feedforward summer is used even though a feedforward multiplier can compensate for the inverse relationship between process gain and flow because of scaling and analyzer reliability issues and the predominant error seen is a bias rather than a span error in the flow measurements. The composition loop trims the feedforward signal. An enhanced PID with a threshold sensitivity setting helps deal with the sample delay and analyzer cycle time and the noise from poor mixing.

Plug flow reactors have a tight residence time distribution but no opportunity for residence time control. The reactor volume must be large enough and the flow distribution uniform enough to provide enough residence time at the highest production rate.

An at-line analyzer and inferential measurement can reduce an excess reactant concentration used to ensure the complete conversion of other reactants.

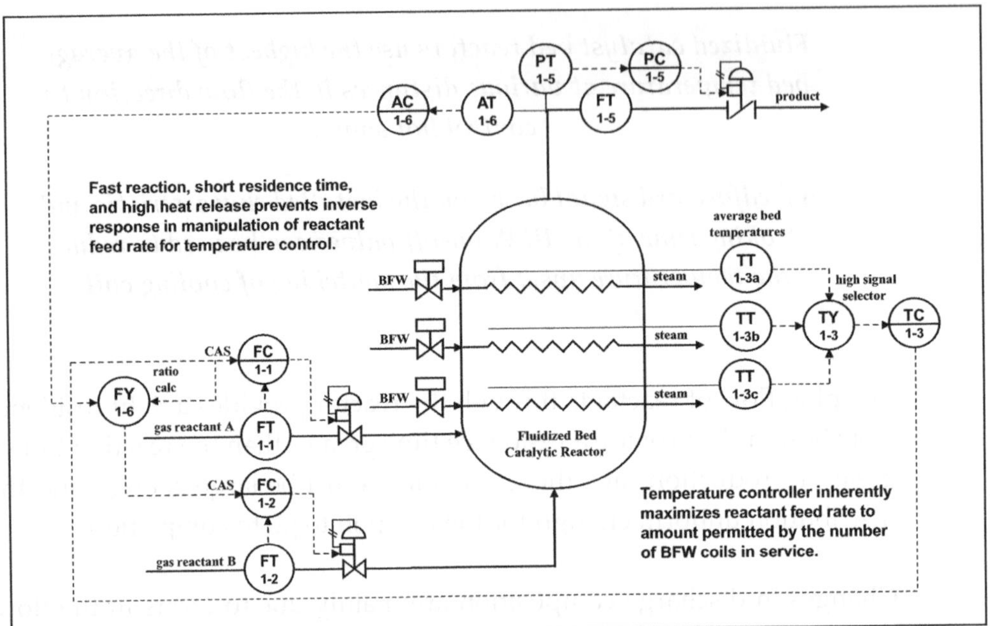

Figure 5.3 – An enhanced PID using an at-line analyzer or inferential measurement for concentration control sets the gas reactant flow ratio, a pressure controller sets reaction time, and a temperature control system maximizes reaction rate by setting gas feed rate

5-2. Mixed Phase Chemical Reactors

Shinskey offers key guidance on reaction completion control in his ISA book *Controlling Multivariable Processes (Ref 82)* that is conceptualized and extended here to include solid phases and batch operations. The term "completion control" is employed instead of the term "endpoint control" used by Shinskey because the term "endpoint" outside of the oil, gas, and petrochemical industry is considered to designate the conditions at the end of a batch operation.

For both batch and continuous reactors, completion control seeks to provide a complete conversion of all the reactants and consequently no excess accumulation of a reactant in a particular phase. If the reactants are in different phases and the product is a single phase, inventory control can be used for reaction completion control. The product must be a gas, liquid, or solids with no recycle or co-products in the other phases.

Inventory control is possible if there is a discharge flow in addition to a feed flow to the phase. Gas volumes have a continuous discharge flow in batch as well as in continuous reactors by means of an off-gas flow going to a vent or recovery system. Liquid and solid volumes only have a discharge flow in continuous operations unless there is a phase separator physically built-in to continuously separate solids from liquids during batch operation (e.g., a hydroclone in a recirculation line).

Completion control seeks to provide for batch and continuous reactors a complete conversion of all reactants.

Using these concepts one can determine whether inventory control provides completion control. For a liquid product, excess gas reactant is inherently prevented by pressure control for both batch and continuous processes. For a gas product, excess liquid reactant is inherently prevented by level control for continuous processes and possibly solids control for all processes.

For multiple phase reactants and a single phase product, inventory control can provide reaction completion control. Batch processes do not have liquid inventory control (level control) to prevent the accumulation of reactants in the liquid phase.

For a liquid product, a pressure controller provides continuous completion control by increasing gas feed for a decrease in pressure from a deficiency of gas reactant and by decreasing gas feed for an increase in pressure from an excess of gas reactant (Figure 5.4). The gas phase reaction is normally fast enough for the gas reactant to be totally consumed in the reaction so that only the inerts are left in the overhead vapor space. An off-gas purge flow prevents the accumulation of inerts. A level controller maintains the liquid material balance by manipulating the liquid product discharge flow.

The residence time control by the level loop shown for a single phase liquid reaction (Figure 5.1a) could help maintain the residence time in the gas phase as well the liquid phase by keeping the bubble rise time constant. The bubble rise time is the bubble path length set by level divided by the bubble velocity set by sparge rate. For pure batch and fed-batch reactors there is no level control and hence no residence time control but otherwise the control scheme is applicable.

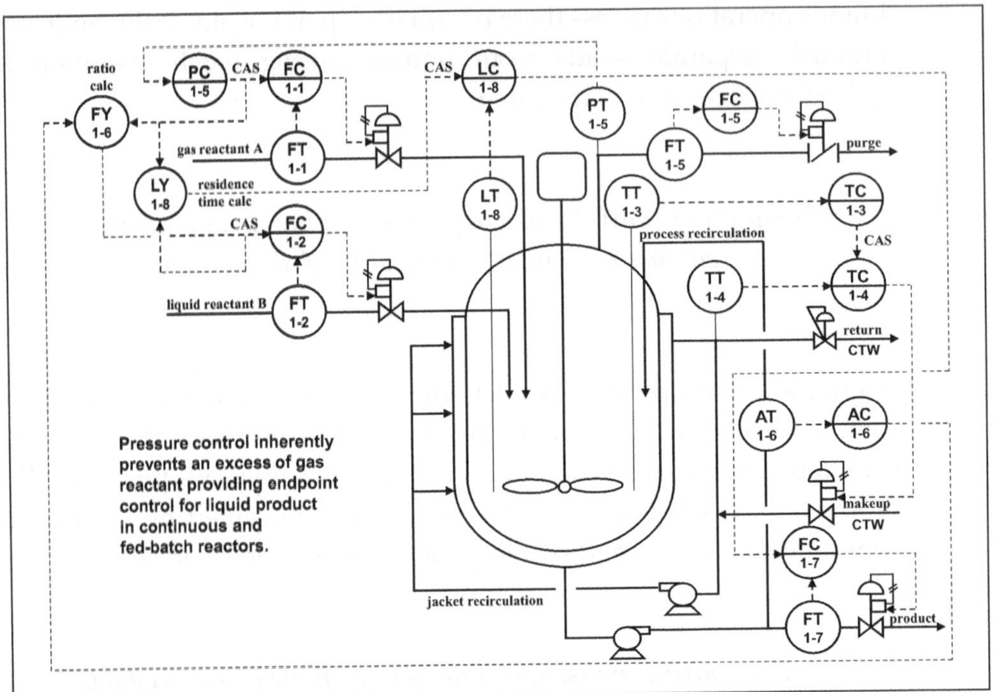

Figure 5.4 – For a liquid and gas reactants, and a liquid product, pressure control maintains continuous composition completion control and level control maintains the liquid inventory

> ***An off-gas purge of inerts enables gas pressure control to provide reaction completion control in the gas phase for a liquid product in batch and continuous reactors.***

For a gas product, a level controller provides continuous completion control by increasing liquid feed for a decrease in level for a deficiency of liquid reactant and decreasing liquid feed for an increase in level from an excess of liquid reactant (Figure 5.5). A purge flow from the bottom prevents the accumulation of inerts in the liquid phase. A gas pressure controller maintains the gas material balance by manipulating the gas product discharge flow. For residence time control, the liquid controller setpoint computed as the liquid reactant flow multiplied by the desired residence time would need a lag inserted because the level controller manipulating the liquid reactant flow forms a positive feedback loop. For fed-batch reactors this strategy is not applicable because there is no level control or purge.

A bottom liquid purge of inerts enables liquid level control to provide reaction completion control in the liquid phase for a gas product in continuous reactors.

Production rate can be maximized in both cases by the use of a valve position controller (VPC) monitoring coolant valve position. The VPC setpoint is the maximum desirable valve position, and the VPC process variable is the temperature controller output. The output of the VPC trims the setpoint of the liquid and gas reactant flow controller for liquid and gas products, respectively. An enhanced PID with dynamic reset limiting for the VPC eliminates limit cycles, reduces interaction between the VPC and temperature controller, and enables smoother optimization with faster correction for disturbances.

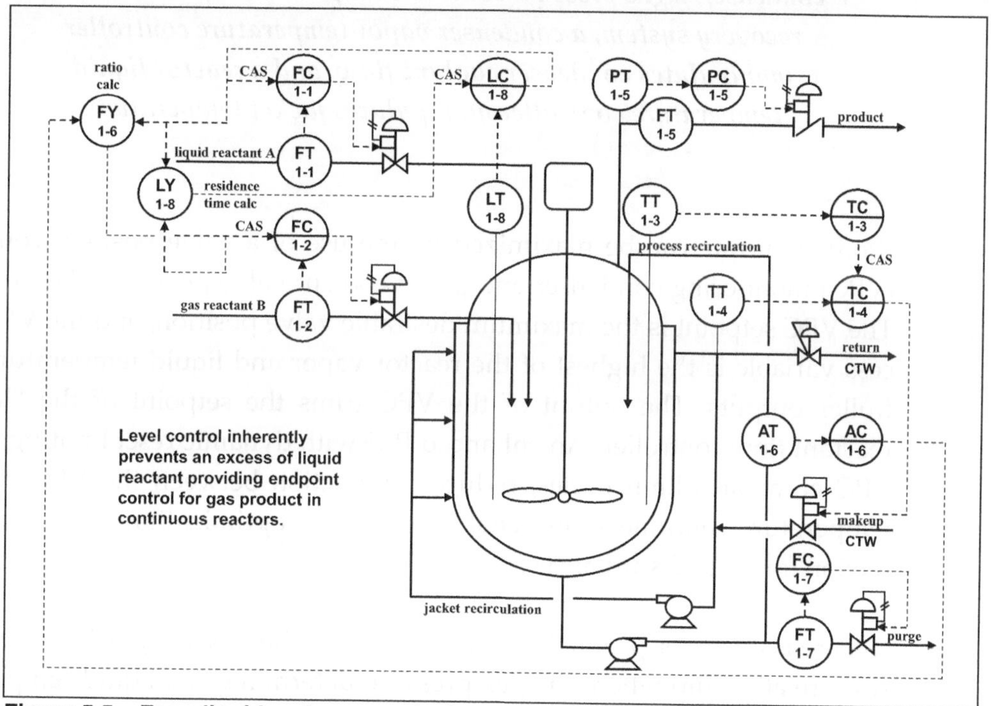

Figure 5.5 – For a liquid and gas reactants, and a gas product, level control maintains continuous composition completion control and pressure control maintains the gas inventory

If there are products in different phases, continuous completion control requires an online or at-line analyzer because the level and pressure controllers have to regulate the liquid and gas, respectively, for inventory control.

Analyzers can provide composition control by manipulating the ratio of reactants or manipulating a two phase product flow.

For a liquid product and a vapor co-product with condensable recycle, a reactor pressure controller manipulates vapor co-product flow. A reactor vapor temperature controller manipulates partial condenser coolant flow and a reactor liquid temperature controller manipulates the secondary jacket coolant temperature setpoint (Figure 5.6a). The reactor level controller manipulates liquid product flow. The level setpoint can be computed to provide residence time control. For fed-batch reactors there is no reactor level control and hence no residence time control but otherwise the same control scheme is applicable.

For liquid and vapor co-product with recycle from a vent condenser, a gas pressure control manipulates vapor flow to a recovery system, a condenser vapor temperature controller manipulates condenser coolant flow, and a reactor liquid temperature controller manipulates jacket temperature.

Production rate can be maximized by the use of a valve position controller (VPC) monitoring condenser and jacket coolant valve positions (Figure 5.6b). The VPC setpoint is the maximum desirable valve position, and the VPC process variable is the highest of the reactor vapor and liquid temperature controller outputs. The output of the VPC trims the setpoint of the "leader" reactant flow controller. An enhanced PID with dynamic reset limiting for the VPC eliminates limit cycles, reduces interaction between the VPC and the temperature controllers, and enables smoother optimization with faster correction for large disturbances.

If there is no vapor product or vapor exit flow, there is a vapor-liquid equilibrium relationship where the gas pressure determines the liquid temperature and vice versa. To eliminate the temperature measurement lag and mixing non-uniformities, an inferential measurement of temperature can be computed from gas pressure using a simple straight line approximation to the pressure versus temperature relationship.

The inferential temperature controller called a P-T controller manipulates the jacket temperature. The P-T controller offers a significantly faster and more

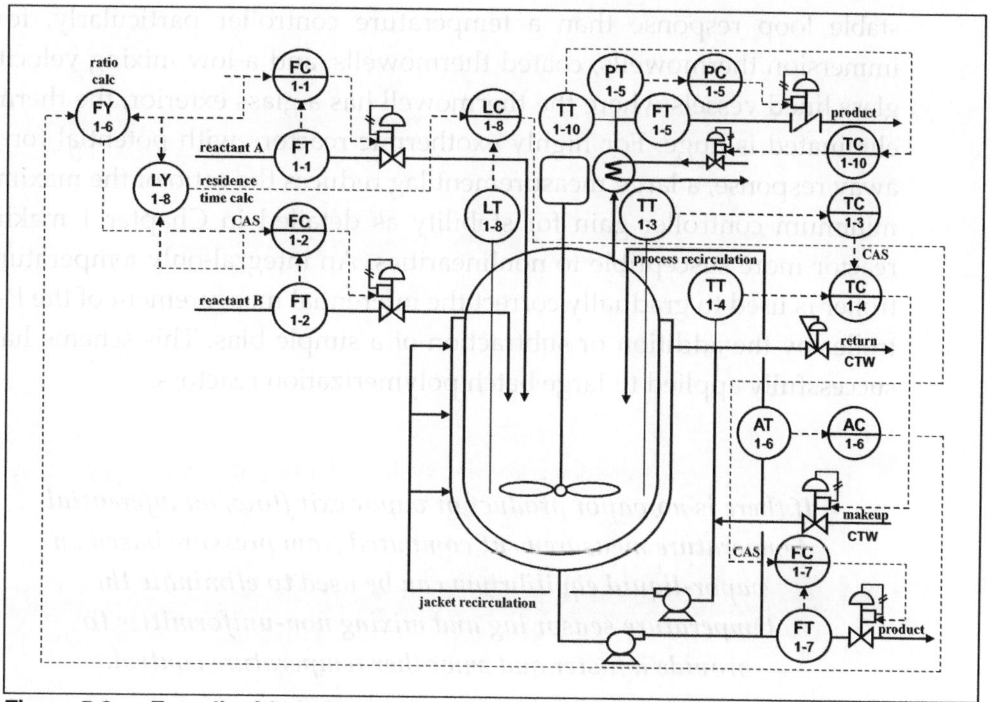

Figure 5.6a – For a liquid product and a gas co-product with condensable recycle and co-product, a reactor pressure controller manipulates vapor product flow

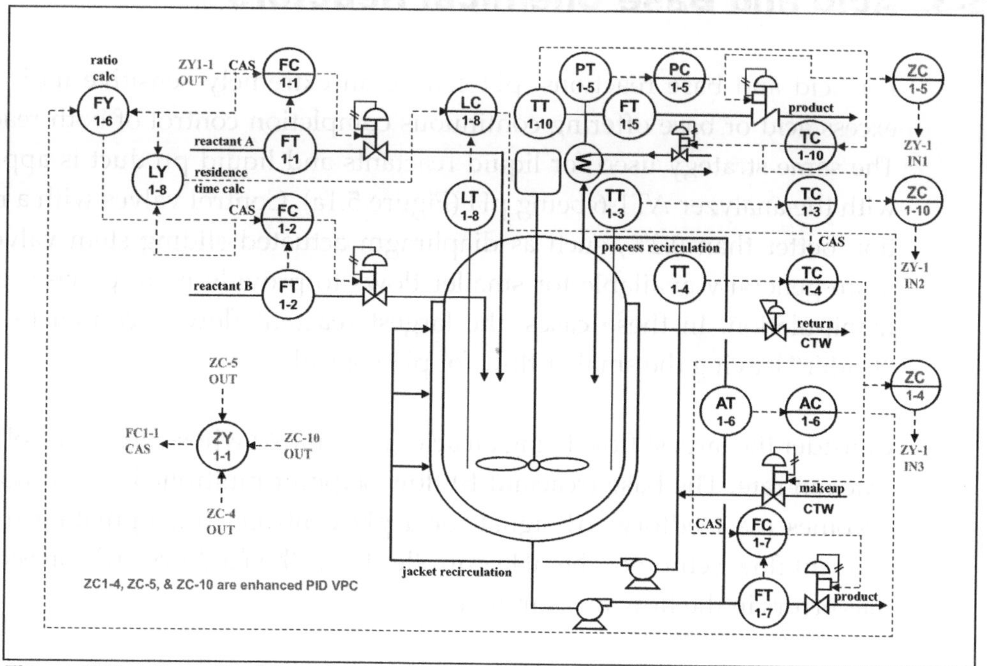

Figure 5.6b – For a liquid product and a gas co-product, the production rate can be maximized by a VPC that increases reactant feed till the coolant valves and gas product valve reach maximum position

stable loop response than a temperature controller particularly, for long immersion thermowells, coated thermowells, and a low mixing velocity. For glass lined vessels where the thermowell has a glass exterior, the thermal lag eliminated is huge. For highly exothermic reactors with potential for a run-away response, a large measurement lag reduces the ratio of the maximum to minimum controller gain for stability as detailed in Chapter 1 making the reactor more susceptible to nonlinearities. An integral-only temperature controller is used to gradually correct the inferential measurement of the P-T controller by the addition or subtraction of a simple bias. This scheme has been successfully applied to large batch polymerization reactors.

If there is no vapor product or vapor exit flow, an inferential temperature measurement computed from pressure based on vapor-liquid equilibrium can be used to eliminate the temperature sensor lag and mixing non-uniformities to provide a faster and smoother temperature control.

5-3. Acid and Base Chemical Reactors

For acid and base reactions, pH can be an extremely sensitive indicator of excess acid or base offering continuous completion control of both reactants. The same strategy used for liquid reactants and liquid product is applicable with the analyzer AT1-6 being pH (Figure 5.1a). Control valves with a resolution better than 0.2%, such as diaphragm actuated sliding stem valves, are more generally available for smaller flows to provide more precise reactant manipulation. In these cases, the largest reactant flow is chosen to be the "leader" leaving the smaller flow for pH control.

Consider the largest flow being an acid (reactant A) whose flow controller sets reaction rate. The base (reactant B) flow setpoint multiplied by a ratio factor becomes the feedforward signal for a pH controller manipulating the acid reactant flow setpoint. The pH controller trims the feedforward correcting for bias errors in the flow measurements.

For acid and base chemical reactors, pH can be a sensitive indicator of excess acid or base concentration and can provide precise reaction completion control.

A reactor level controller manipulates discharge product flow to control inventory and residence time. For fed-batch reactors there is no reactor level control and hence no residence time control but otherwise the same control scheme is applicable.

Production rate can be maximized by the use of a valve position controller (VPC) monitoring acid or base reactant valve positions just as depicted for the liquid chemical reactor (Figure 5.1b). The VPC setpoint is the maximum desirable valve position, and the VPC process variable is the temperature controller output. The output of the VPC trims the setpoint of the leader reactant flow controller. An enhanced PID with dynamic reset limiting for the VPC eliminates limit cycles, reduces interaction between the VPC and the temperature controller, and enables smoother optimization with faster correction for disturbances.

5-4. Yeast Bioreactors (Fermenters)

Bioreactors that use yeast are called fermenters and the liquid is referred to as broth. This section focuses on the control and optimization of front end of ethanol plants but concludes with an overview of fermenters for alcoholic beverage production.

Yeast fermenters to produce ethanol traditionally only have temperature control. The subscribed permissible pH range is quite large (e.g., 4 to 6 pH). These fermenters typically don't have pH control. A gradual decrease in pH during ethanol fermentation is normal due to acetic acid and lactic acid formation. A greater than normal drop in pH is indicative of a bacterial contamination so pH may be monitored. The main opportunity for maximizing production is taking advantage of a decrease in the time to reach the desired alcohol batch concentration endpoint. The inhibition effect of alcohol concentrations above the endpoint dramatically slows down and suspends further conversion to alcohol. Consequently, longer batch times do not translate into greater ethanol concentrations.

Production rate can be increased by reducing the batch time to when the endpoint is reached. An online near infrared analyzer or an at-line chromatograph can provide a measurement of ethanol concentration. Most batch cycle times are much longer than the average time to an endpoint because of the variability in time to the endpoint. If batch cycle times are fixed, a consistent early endpoint can be used to reduce corn feed rate to improve ethanol yield, which is typically expressed as gallons per bushel. Corn is the largest cost for ethanol production. An improvement in yield taken as a decrease in corn feed rate significantly decreases cost and decreases the carbon footprint, which increases revenue.

The front end of the ethanol plant consists of slurry, liquefaction, cooling, and saccharification-fermentation (SSF) areas. The front end is continuous with batch operation of the fermenters in the SSF area. The basic control system for the front end of an ethanol plant controls the temperature and pH of the fermenter feed. The fermenter may have temperature control as well.

As previously mentioned, the pH in the fermenter will fall over the course of the batch from acetic acid and lactic acid formation. While there is an optimal pH for yeast growth and product formation, the effect of the change in fermentation batch cycle time is not thought to be significant enough to warrant pH control in the fermenter.

The control system developed for optimization of ethanol yield in the front end uses three enhanced PID controllers, an at-line corn feed analyzer, and an at-line multiplexed ethanol analyzer for the fermenters (Figure 5.7). The PID controllers optimize the corn feed rate and solids in the slurry tanks based on predicted fermentable starch and actual SSF batch time to reach the ethanol endpoint. A virtual plant using first principle models of the front end of an ethanol plant is used to develop and prototype the optimization strategies.

The first enhanced PID AC1-4 controller is on the first slurry tank (Figure 5.7). The PID is a production rate controller with its PV and SP scaled to a 0-400 gpm ethanol production rate. The PV is computed based on corn feed rate and the percent of fermentable solids, with an inferential prediction of ethanol yield being provided by an at-line analyzer taking samples from the corn feeder. External reset feedback (dynamic reset limit) is enabled so that the PID does not try to drive its output faster than the corn feeder can respond. The output of the PID manipulates corn feeder speed. For an increase in the infer-

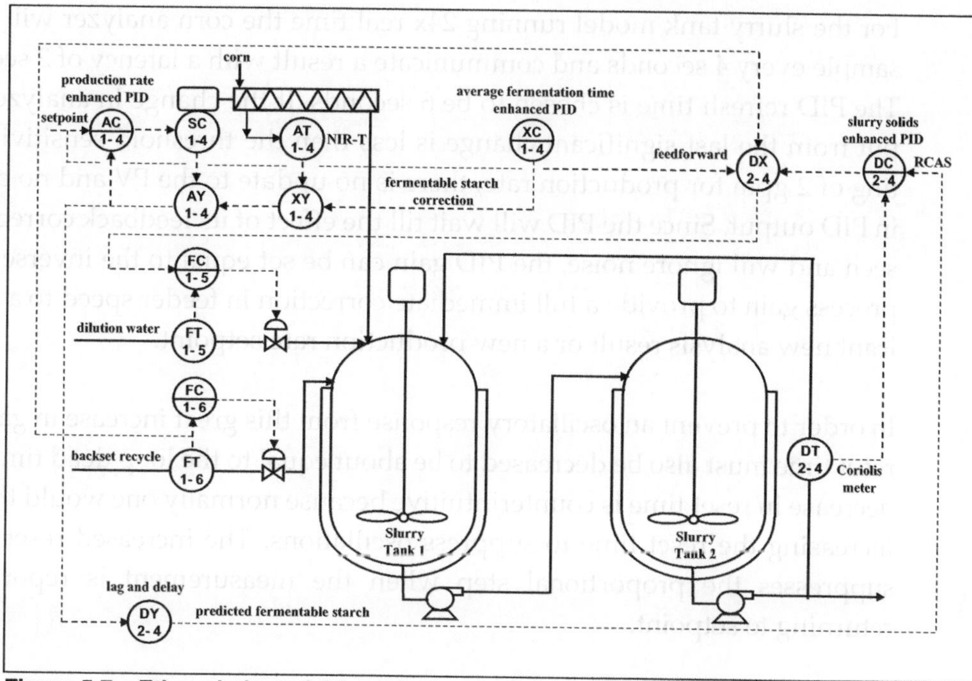

Figure 5.7 – Ethanol plant yield can be increased by the use of an at-line corn feed analyzer and enhanced PID to optimize corn feed rate, slurry solids concentration, and batch time

> ***Enhanced PID controllers can optimize ethanol plant yield by the use of at-line analyzers.***

ential prediction of yield, the PID will immediately cut back on corn feed rate, reducing corn use. The simple change of the PID setpoint enables operations personnel to change front end production to match the separation and purification capability of the back end.

> ***An enhanced PID uses as its controlled variable a front end production rate prediction computed from corn feeder speed and an at-line corn analyzer prediction of yield. The PID can immediately translate an improvement in corn yield capability to a reduction in corn use.***

For the slurry tank model running 24x real time the corn analyzer will pull a sample every 4 seconds and communicate a result with a latency of 2 seconds. The PID refresh time is chosen to be 6 seconds. If the change in analyzer output from the last significant change is less than the threshold sensitivity setting of 2 gpm for production rate, there is no update to the PV and no change in PID output. Since the PID will wait till the effect of its feedback correction is seen and will ignore noise, the PID gain can be set equal to the inverse of the process gain to provide a full immediate correction in feeder speed to a significant new analysis result or a new production rate setpoint.

In order to prevent an oscillatory response from this great increase in gain, the reset time must also be decreased to be about equal to the loop dead time. This decrease in reset time is counterintuitive because normally one would think of increasing the reset time to suppress oscillations. The increased reset action suppresses the proportional step when the measurement is reported as returning to setpoint.

Changes in analyzer cycle time or communication latency do not affect the tuning because PID action is suspended until there is an update. For changes in predicted yield (fermentable starch), the PID makes a single adjustment in corn feed rate bringing the production rate back to setpoint. The figure also shows that production setpoint changes can be easily made to match corn supply, market demands and back end capacity constraints.

> ***An ethanol plant production rate controller can make a single correction for an at-line analyzer update of a change in corn yield that will return the production rate to setpoint. The immediate and nearly exact correction is the result of being able to set the enhanced PID gain to be the inverse of the open loop self-regulating process gain.***

The second enhanced PID DC2-4 controller is on the second slurry tank (Figure 5.7). This PID is a slurry solids concentration controller with its PV and SP scaled to be 0-40% weight percent fermentable solids. Slurry solids concentration is inferred from a Coriolis meter density measurement in a recycle stream. The output of the PID manipulates the dilution water to the first slurry tank. Since the manipulation is to the first slurry tank while the solids being controlled are in the second slurry tank, there is an equivalent dead time from the residence time of the two slurry tanks plus mixing and injection delays.

The total loop dead time measured online is about 80 seconds in the model. A feedforward signal is computed for the dilution water based on corn feed rate, steam injection, and backset recycle. To ignore feedforward timing errors, an update time of 12 seconds is set on the inferential measurement of fermentable solids. A dead time block with a dead time parameter of 80 seconds in the external feedback path from the analog output to the PID provides dead time compensation. A threshold sensitivity setting is used to ignore measurement noise.

Since the PID compensates for dead time, ignores noise, and waits out feedforward timing errors, the PID gain can be increased to provide a faster setpoint response. The feedforward signal from the change in Slurry Tank 1 prevents the changes in production rate from affecting the solids concentration in Slurry Tank 2. The feedforward signal is the production rate setpoint minus the water flow from other sources such as backset. The remote cascade (RCAS) setpoint is the desired solids concentration multiplied by a fermentable starch factor predicted by the corn analyzer. The prediction passes through a delay and lag set equal to the dead time and residence time, respectively, so the change in setpoint coincides with the change in solids measurement from a change in corn feeder speed from the production rate controller.

An enhanced PID for slurry solids control in the front end of an ethanol plant is improved by addition of dead time compensation, a feedforward of dilution water and steam injection rate, and a correction of setpoint based on corn feeder speed. The PID uses a Coriolis meter in a slurry tank recirculation line to provide an inferential measurement of percent solids.

The third enhanced PID XC1-4 controller (Figure 5.7) uses a High Performance Liquid Chromatograph (HPLC) measurement of ethanol profiles in simultaneous saccharification and fermentation (SSF) batches. Saccharification uses an enzymatic reaction to convert maltose, maltriose, and higher polymer dextrins to glucose. Fermentation employs yeast to produce ethanol from glucose. In older plants, the saccharification and fermentation are done in separate vessels.

SSF decreases the total batch cycle time. The HPLC provides an inferred measurement that is a running average of the fermentation time (batch time to ethanol endpoint) in the most recent SSF batches. The model has multiple SSF vessels each with a saccharification-fermentation time of 8 minutes (240x speedup), 80 seconds (24x speedup) for charging, and 40 seconds (24x speedup) for draining and preparing for the next batch. When the HPLC indicates that the ethanol has reached the desired endpoint, the running average of batch times is updated. Note that the actual batch cycle time is fixed. The average is the time to the desired endpoint.

After an initial delay of 10 minutes, a new batch time result is available and the running average is updated every 2 minutes in the model. The update is faster for a decrease in batch time from an increase in actual fermentable solids that can go from sugars to ethanol in the SSF batch. The enhanced PID output biases a correction to analyzer measurement of predicted fermentable solids, which changes the inferred production rate. The production rate controller correspondingly adjusts the corn feeder speed to bring the production rate back to setpoint. An unmeasured increase in fermentable solids will show up as an early achievement of the batch endpoint resulting in a cutback in corn feed rate and providing an immediate improvement in yield.

An enhanced PID controller can gradually correct the inferential prediction of ethanol yield used by the front end production rate controller by averaging the time that fermenter batches take to reach the desired ethanol endpoint.

In beer, wine, and other alcoholic beverage production, the mixture of fermentable sugars and other ingredients developed by brew masters from taste tests is critical. The masters are not particularly interested in more sophisticated process control. Agitation may not be used. Special fermenter wall material (e.g., oak) is used to help establish flavor. For small-volume specialty beverages, the mixture may just sit in a barrel in a cool room. For large scale production fermenters with specific wall designs (e.g., stainless steel dimpled jackets) are used for more efficient and repeatable operation.

For the big producers, off-line lab analysis is used for monitoring beverage consistency. Production can be increased by knowledge of when fermentation is complete. Carbon dioxide production rate (CPR) can be used as an inferential measurement of alcohol formation rate. Load cells and a simple loss in weight calculation have been used for wine fermenters to provide an online CPR and alcohol formation rate since there is no flow into the batch (no sparger or reagents) and the only flow out of the batch is the carbon dioxide exiting as vent gas. The loss in weight calculation uses a dead time block discussed for batch profile control (Section 5.8) to provide fast updates with a good signal to noise ratio.

For large-scale commercial wine production, an inferential measurement of alcohol formation rate from a loss in weight calculation from load cells can monitor production consistency and increase production rate by providing knowledge of when the fermentation is complete.

New processes are being developed to use genetically engineered yeast to produce large volume chemical intermediates, such as adipic acid that have been traditionally produced by petrochemical processes. The process control requirements are extensive for these new products.

5-5. Fungal Bioreactors

Fungal bioreactors have been extensively employed to produce a wide spectrum of antibiotics and some chemical intermediates such as citric acid. In addition to temperature control, these bioreactors have air flow control, agitation control, dissolved oxygen (DO) control, pH control, and pressure control. Dissolved oxygen control may be split ranged to manipulate agitation rate and air flow. The split range may be extended to include vessel pressure if higher pressures do not cause undesirable higher dissolved carbon dioxide concentrations. The pH control may be split ranged between an acid reagent (e.g., sulfuric) and a base reagent (e.g., ammonia). Glucose addition rate is fixed or scheduled to provide a more constant concentration. Many of the opportunities for increasing production rate by modeling and advanced control discussed for mammalian bioreactors (Section 5.7) are applicable to fungal bioreactors.

5-6. Bacterial Bioreactors

Bacterial bioreactors are used to produce more complex molecules such as the human epidermal growth factor (hEGF). Bacterial bioreactors have the same type of controls as the fungal reactors. The split range for dissolved oxygen (DO) control may be extended from simple air flow control to include pressure control and agitator speed control. The rangeability requirement for oxygen demand is addressed by low agitation and low air flow in the pre-exponential growth phase when oxygen demand is low. At this point, the DO control may be noisy due to intermittent bubbles from lower sparger flow rates and less mixing. Many of the opportunities for increasing production rate by modeling and advanced control discussed for mammalian bioreactors are applicable to bacterial bioreactors. Oxygen uptake rate (OUR) from mass spectrometer measurements can be used as an inferential measurement of cell growth rate.

> *Bacterial bioreactors meet the extreme rangeability of oxygen uptake rate by a split range between vessel pressure control, agitator speed control, and air flow control.*
>
> *Oxygen uptake rate (OUR) can be computed from mass spectrometer measurements to provide an inferential measurement of cell growth rate.*

5-7. Mammalian Bioreactors

Mammalian bioreactors have the most stringent control requirements particularly for pH and temperature. The liquid termed media consists of water, nutrients, and a substrate of sugar (glucose) and amino acid (glutamine). Agitation is kept at a low rate because mammalian cells have no cell walls (they have cell membranes) and are easily damaged by impeller shear. Mammalian bioreactors have all of the process control for bacterial bioreactors but with more complex DO and pH loops (Figure 5.8a). Sparger design is critical to provide dispersion of gases and mixing and reduce interaction between the DO and pH controllers.

In an effort to increase the safety, efficiency and affordability of medicines, the Food and Drug Administration (FDA) has proposed Process Analytical Technology (PAT) guidance to encourage innovation and improvement in pharmaceutical development, manufacturing, and quality assurance. The primary focus of the PAT guidance is to reduce variability through a better understanding of a process than can be obtained by the traditional approach. PAT is formally defined by the Food and Drug Administration (FDA) in Part IV of the PAT guidance as follows:

"The Agency considers PAT to be a system for designing, analyzing, and controlling manufacturing through timely measurements (i.e., during processing) of critical quality and performance attributes of raw and in-process materials and processes, with the goal of ensuring final product quality. It is important to note that the term analytical in PAT is viewed broadly to include chemical, physical, microbiological, mathematical, and risk analysis conducted in an integrated manner. The goal of PAT is to enhance understanding and control the manufacturing process, which is consistent with our current drug quality system: quality cannot be tested into products; it should be built-in or should be by design."

While analyzers are the key to better process understanding and control, PAT is more than just about adding analyzers. PAT seeks to promote better use of the data. The term "analytical" refers the ultimate goal of developing knowledge of the nature, proportion, function, and interrelationship between the constituents of the manufacturing process.

The DO controller output is split ranged to manipulate air and oxygen sparge and air overlay (vapor space addition). The pH controller output is split ranged between carbon dioxide addition (carbonic acid) and sodium bicarbonate (base). These are weak acids and bases, providing a great moderating (buffering) effect on the titration curve slope and enabling much tighter pH control.

Weak acids and bases, and the use of a tiny aperture for the reference electrode junction, reduce the liquid junction potential providing a more accurate

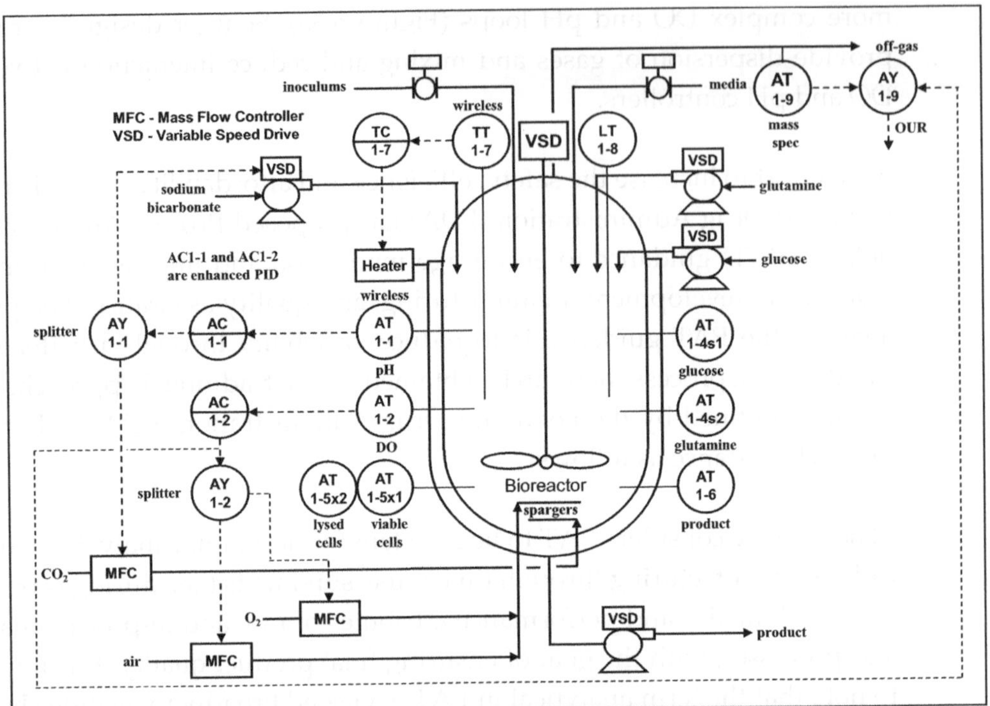

Figure 5.8a – Mammalian bioreactors have all of the process control of bacterial reactors but with more complex DO and pH loops with decoupling and smarter split ranged control

> ***Mammalian cell bioreactors have more stringent control requirements and a greater opportunity for PAT initiatives and process optimization due to the sensitivity of the cells to operating conditions and sophistication of the cell functions.***

pH measurement. The disturbances from changes in cell concentration and function take days to develop (slower than concentration disturbances in any other type of reactor). With proper tuning, pH control within 0.01 pH is possible. Instead of dealing with cell disturbances, the focus is on eliminating overshoot for changes in pH setpoint, reducing interaction with the DO control, and preventing high carbon dioxide concentrations that can suffocate cells and the excessive accumulation of sodium bicarbonate concentrations that can increase the osmotic pressure (cell internal pressure).

Incredibly tight pH control is possible because of buffering and extremely slow disturbances.

Interaction between DO and pH control can be greatly reduced by sparger design. Remaining interaction can be dealt with by an enhanced PID for DO and pH control with a threshold sensitivity setting for ignoring noise and a decoupling feedforward signal that is the opposing controller's gas addition rate. Since tight pH control is more important than DO control, a feedforward of DO PID output (e.g., air flow setpoint) can be added to the pH PID output to help the pH controller immediately correct the carbon dioxide (CO_2) flow setpoint.

The feedforward does not need to correct the pH PID output when manipulating the sodium bicarbonate setpoint. This simple half decoupler of air to CO_2 flow setpoint is usually sufficient unless there is a sparger design problem. Problems occur when air and CO_2 flow share the same sparger, the spargers are too close to each other, or there is no tight pressure control of air or CO_2.

Good sparger design can eliminate most of the interaction between pH and DO control. What remains can be reduced by a half decoupler where the air flow setpoint from the DO PID output is added to the pH PID output that is the CO_2 flow setpoint.

High concentrations of carbon dioxide (CO_2) can be avoided by preventing overreaction of the pH controller to pH shifts (pH setpoint changes to promote product formation) by the use an enhanced PID and a secondary CO_2 flow PID with setpoint rate limits. High CO_2 concentrations will dissipate with time. Inert gases are sometimes added to help sweep CO_2 out of the batch.

Mammalian bioreactor batches take 10 or more days. The exceptionally slow kinetics translate to extremely slow changes in reagent, oxygen, and cooling demand. Exceptionally slow load disturbances and slow integrating process gains result in negligible errors from disturbances. The objectives of tuning and control loop performance for pH, dissolved oxygen, and temperature control are in terms of setpoint response and optimization rather than disturbance rejection. The recognized opportunity is getting to a setpoint with no overshoot for a temperature shift and pH shift (setpoint change) to move from enhancing cell growth rate to promoting product formation rate. A PID structure that has proportional action on the process variable rather than on error and no derivative is useful for preventing overshoot. Getting to setpoint faster is not a consideration since the minutes saved are inconsequential in terms of the batch cycle time.

Mammalian cell kinetics are extremely slow with batch cycle times of 10 or more days. The errors from load disturbances are so slow that tuning for disturbance rejection or getting to setpoint faster is not the goal. Most of the focus is on the prevention of overshoot.

As previously mentioned, excessive accumulation of sodium bicarbonate causes a high sodium concentration leading to a high osmotic pressure that can contribute to cell membrane rupture and cell death. Unlike carbon dioxide that can escape in the vent gas, sodium bicarbonate is trapped in the media. Excessive crossing back and forth of the split range point for pH control over the course of the batch can lead to high osmotic pressure. A split range gap or deadband complicates tuning and does not stop the eventual excursion from the integrating process response. External reset feedback (dynamic reset limit) in the primary pH PID and setpoint rate limits in the secondary flow PID can provide the directional move suppression needed to slow down a movement into the direction of adding sodium bicarbonate, reducing unnecessary split range excursions without requiring special tuning. New at-line analyzers can measure osmolality (osmotic pressure) by freeze point depression to help in optimization of the setpoint rate limits.

New online and at-line analyzers for measuring glucose and glutamine open the opportunity for improved concentration control of this sugar and this amino acid. At present, glucose and glutamine addition are scheduled in anticipation of utilization rates. A move from sequenced batch charges to fed-batch closed loop control can compensate for disturbances and consistently maintain a concentration (Figure 5.8b). The glutamine feed setpoint can be

External reset feedback in the pH PID and setpoint rate limits in the CO_2 and the sodium bicarbonate flow controllers can provide directional move suppression to eliminate unnecessary crossings of the split range point and an increase in osmotic pressure.

ratioed to the glucose feed setpoint. A feedforward signal for glucose addition rate based on oxygen uptake rate (OUR) can provide preemptive correction for the acceleration of substrate utilization in the exponential growth phase. The sampling rate for at-line analyzers doesn't need to be more frequent than once per 4 hours because the changes in mammalian cell metabolism are slow. An enhanced PID can provide tight control for relatively large cycle times and can suppress reaction to noise and feedforward timing errors by a threshold sensitivity setting.

New online and at-line analyzers of glucose and glutamine concentration offer the opportunity for optimizing the ratio of these concentrations for better cell performance (higher cell growth rate and product formation rate and better product quality).

5-8. Biological Fed-Batch Profile Control

If the rate of change is computed for viable cell and product concentration by passing these concentrations through a dead time block to create an old value that is subtracted from the new value and divided by the dead time, the cell growth rate and product formation rate can be indicated and controlled (Figure 5.8c). The concentrations can be measurements from online or at-line analyzers or inferential measurements from OUR or a virtual plant running in synch with the actual plant. The batch profile for both rates tends to start slowly, accelerate, and then decline.

Profile control can provide a more consistent and faster rate (e.g., cell growth rate and product formation rate) at the most appropriate time. The glucose setpoint or the glutamine to glucose ratio may be directly manipulated by cell

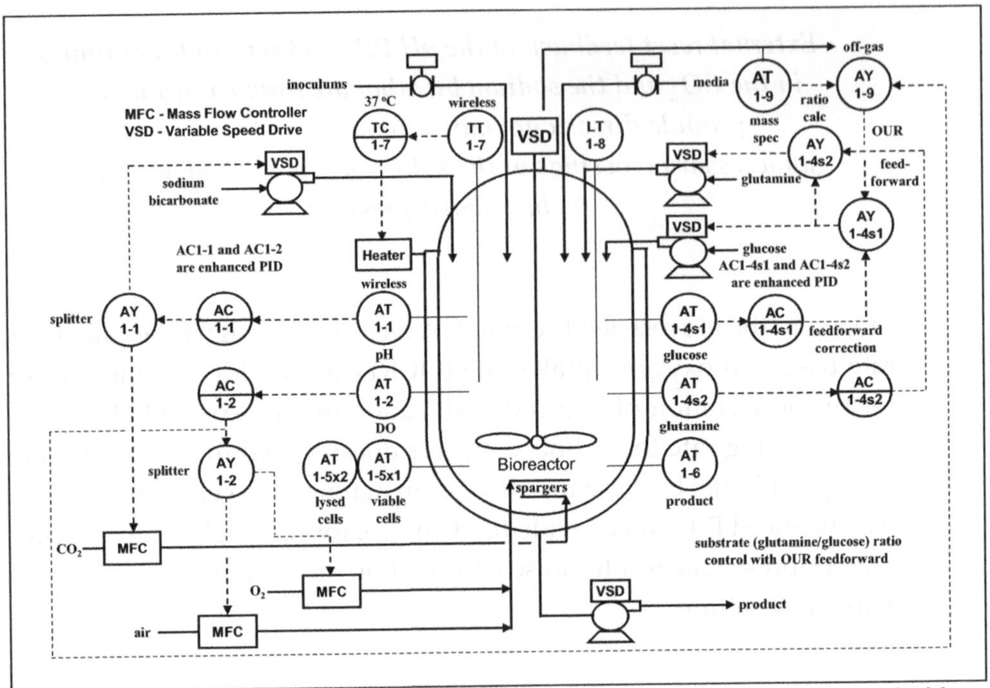

Figure 5.8b – Online and at-line analyzers enable substrate concentration control with glutamine feed ratioed to glucose. An OUR feedforward anticipates changes in glucose utilization rate

growth rate or product formation rate controllers whose setpoint is a function of batch phase. Product formation rate control may start later in the batch at which point growth rate control may be suspended or given less priority through tuning. An enhanced PID is used for these controllers to deal with discontinuous measurements, noise, and interactions.

The manipulation of temperature or pH as a secondary control loop for profile control requires a translation of the controlled variable from rate of change (velocity) to acceleration and deceleration to deal with the peak in the plot of growth rate or formation rate versus temperature and pH (Figure 4.1a and 4.1b). The change in sign of the slope and process gain would confuse a controller whose process variable is the slope. The creation of a new controlled variable for the secondary loop that is the change in batch profile slope with respect to a change in temperature or pH eliminates the reversal of the sign of the process gain. The controlled variable decreases as the operating point approaches the optimum in the plot of cell growth rate or product formation rate versus pH or temperature. The controlled variable is zero at the peak and becomes more negative as the operating point moves further to the right of the peak.

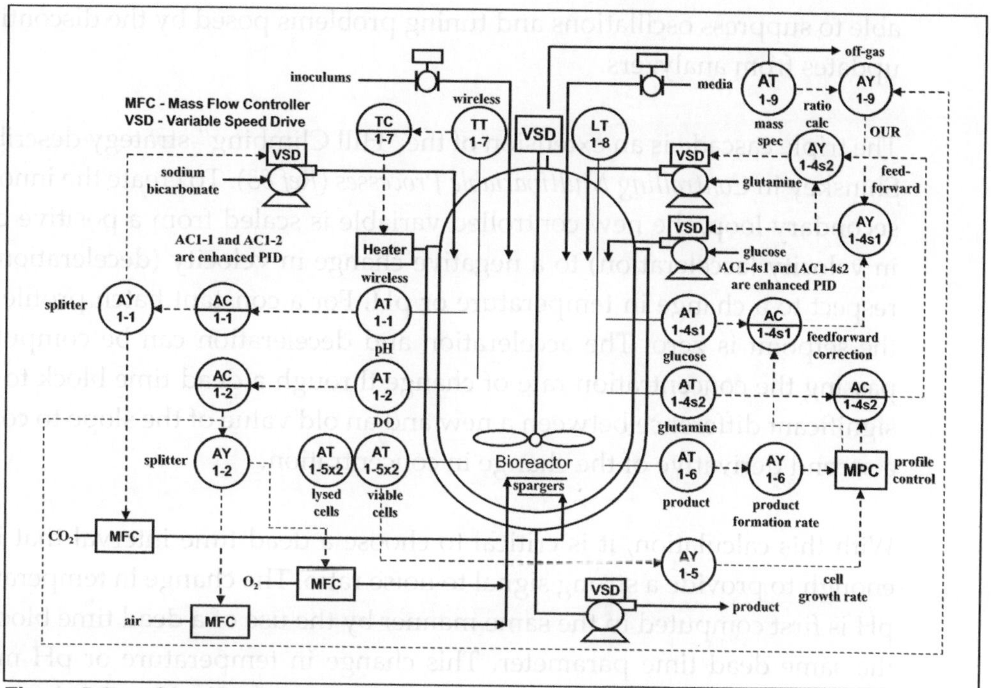

Figure 5.8c – Growth rate and product formation rate from the rate of change of actual or inferential measurements can provide fed-batch profile control by the manipulation of glucose and glutamine

> ***An enhanced PID, and a dead time block computation of the rate of change of a concentration, create the opportunity for batch profile slope control to provide a more consistent and optimized cell growth rate and product formation rate.***

The manipulated variable for this new secondary controller is the temperature or pH setpoint of a tertiary lower loop. Thus, there is a triple cascade of batch profile slope to batch profile acceleration to temperature or pH. However, unless Model Predictive Control (MPC) is used, the manipulation of either temperature or pH may be done and not both, due to interactions. In addition, careful setpoint limits must be placed on the tertiary controller. If PID controllers are used, external reset feedback throughout the triple cascade is critical to prevent an upper PID output from changing faster than a lower PID process variable can respond. External reset feedback helps deal with the violation of the cascade rule caused by the secondary loop not being 5 times faster than the primary loop. The enhanced PID described in Appendix E is advisable to suppress oscillations and tuning problems posed by the discontinuous updates from analyzers.

The triple cascade is an extension of the "Hill Climbing" strategy described by Shinskey in *Controlling Multivariable Processes* (*Ref* 83). To create the innovative secondary loop, the new controlled variable is scaled from a positive change in velocity (acceleration) to a negative change in velocity (deceleration) with respect to a change in temperature or pH. For a constant batch profile slope, the setpoint is zero. The acceleration and deceleration can be computed by passing the concentration rate of change through a dead time block to give a significant difference between a new and an old value of the slope to compute a second derivative of the change in concentration.

With this calculation, it is critical to choose a dead time interval that is long enough to provide a strong signal to noise ratio. The change in temperature or pH is first computed in the same manner by the use of a dead time block with the same dead time parameter. This change in temperature or pH must be dynamically compensated so that a change in profile is coincident with a change in temperature or pH. The dynamic compensation can be identified by the use of MPC software similar to what was described in Section 2.4 for linear dynamic estimators except that the time intervals between step tests must be larger than the dead time block parameter used to create the rate of change signals. The calculation must be protected against a divide by zero.

When the triple cascade is turned on, a change in the manipulated temperature or pH in the most desirable direction should be initiated to create a change in slope. The tuning of the primary PID must be slower than that of the secondary PID, which must be slower than that of the tertiary PID. Since this is an optimization, and biological batch cycle times are extremely long, patient, gradual corrections are in order by the use of exceptionally large reset times and low PID gains. The product of the PID gain and reset time must be greater than the inverse of the integrating process gain for the primary and secondary loops.

Smooth and gradual action is also important from the standpoint of not reacting to noise. The numerator used in the calculation of the controlled variable for the secondary PID is a second derivative of the batch profile with respect to time giving the acceleration or deceleration in the batch profile with respect to a change in pH or temperature. The computation is consequently susceptible to noise.

For maximizing the cell growth rate or product formation rate, a triple cascade extension of the Shinskey "Hill Climbing" technique can optimize the pH or temperature setpoint to be near the peak in the plot of cell growth rate or product formation rate versus pH or temperature. For temperature, the target is just to the left of the peak because of the risk of cell death from temperatures to the right of the peak from sensor offset or setpoint overshoot.

6
Cooling and Heating Control Systems

6
Cooling and Heating Control Systems

The heating and cooling system is critical for reactors since temperature plays such a huge role in determining reaction rate and selectivity (formation of desired product versus byproducts or waste). The limit to what a valve position controller (VPC) can do to increase production rate depends upon the capability of the heating and cooling system. The system can also be the source of discontinuities and nonlinearities. Most reactor temperature control problems beyond tuning can be traced back to deficiencies in the reactor coil or jacket temperature, mechanical, process, piping and control system design.

Heating is required for endothermic reactions (reactions that consume energy), for vaporizing liquids, or for bringing a reactor up to operating temperature. Cooling is needed for exothermic reactions (reactions that product heat), condensing vapors, and to bring a reactor down to operating temperature. Both heating and cooling are often required. Split ranged control is used to go back and forth between heating and cooling. In split ranged control for an exothermic reactor with a fail open cooling valve, as the temperature controller output increases from 0% to the split range point, the coolant valve goes from wide open to completely closed. As the temperature controller output increases from the split range point to the 100%, the heating valve goes from closed to wide open. The split range point is traditionally set at 50% but would be better set so that the change in temperature or rate of change of temperature for a change in flow is about the same for each valve (e.g., open loop process gains are the same for each valve).

Cooling and heating can be applied directly to the reactor by heat transfer surfaces in the vessel exterior (e.g., jackets) or interior (e.g., coils) or indirectly via a process recirculation stream. Heat exchangers are used for some jacket temperature control systems and for all process recirculation temperature control systems.

6-1. Jacket and Coil Temperature Control

Coils offer faster and more aggressive heating and cooling. Jackets offer a more even heat transfer distribution. However, jackets have a lower utility velocity at the heat transfer surface that can increase fouling rate and require good mixing to ensure that the jacket has a uniform effect on the entire liquid contents. Coils offer excellent heat transfer to fluids in the interior but generally not in a dished bottom. At low liquid levels, the process liquid may not be in contact with the coils. For highly exothermic plug flow reactors such as gas reactors, coils provide more effective cooling from direct process contact in the core and high fluid velocities inside and outside the coils.

> ***Coils offer faster and more aggressive cooling and do not require a well-mixed process fluid but do not offer good heat transfer for low liquid levels.***

Many of the temperature control considerations for coils and jackets are the same. For example, both coil and jacket systems benefit from keeping the utility flow through the coil and jacket constant by the use of a recirculation pump and piping to return utility flow from the outlet back to the inlet of the coil or jacket, and from eliminating split range discontinuities by control valve and control system design.

> ***Coils and jackets both benefit from a recirculating utility flow and better split range control.***

A constant jacket flow provided by recirculation and the throttling of a makeup flow is preferred to the throttling of total flow to the jacket. The process gain and dead time for the jacket temperature controller are inversely proportional to flow. At low jacket flows the combination of high process gain and high process dead time can cause sustained oscillations. In addition, at low jacket flows the heat transfer coefficient is reduced and the fouling rate is greatly increased, decreasing the capability of the reactor temperature control system to do its job.

Throttling a utility makeup flow instead of the jacket or coil flow helps to keep the jacket or coil dynamics constant and reduces fouling of the utility side of heat transfer surfaces.

Mechanical design needs to prevent mixed phases in the jacket or coil (e.g., steam bubbles in coolant and coolant droplets in steam) after a transition between cooling and heating. Control valves must be precise (e.g., minimum backlash and stick-slip) and properly sized to prevent limit cycles and abrupt changes as one valve closes and another valve opens. Valves designed for tight shutoff have excessive seating and sealing friction near the closed position. Isolation valves are not throttling valves and vice versa. To prevent leakage of a utility flow, an isolation valve should be added in series with the throttling valve to open and close on an on-off basis as the throttling valve strokes open or closed.

Mechanical design and control valve design should prevent mixed phases and abrupt changes in heat transfer medium, temperature or flow rate at a transition between heating and cooling.

All of the control schemes can benefit from the use of external reset feedback (dynamic reset limit) in the reactor and jacket or coil PID controllers. External reset feedback can prevent the primary reactor temperature PID from trying to change the secondary temperature PID setpoint faster than the jacket or coil temperature can respond. External reset feedback with a fast utility valve position readback can help prevent limit cycles from valve backlash that are particularly problematic near the split range point. The reset time can adapted to reduce the overshoot by a providing a smarter time for the valves coming off output limits for large setpoint changes in batch operation and continuous operation transitions and startups.

For highly exothermic reactors, external reset feedback offers the ability to reduce the chance of a runaway by the addition of directional rate setpoint limits to provide a faster response to a demand for cooling. Setpoint rate limits on the analog output block to valves or on the PID for a secondary flow loop provide directional move suppression to help reduce unnecessary crossings of the split range point. The setpoint rate limit is slower in the direction of moving to the split range point. External reset feedback of the actual response of the valve or secondary flow is used to prevent the jacket or coil temperature PID output from trying to change faster than the setpoint rate limits will allow.

> ***External reset feedback can help eliminate most sources of oscillations from cascade control and split ranged valves.***

Cascade temperature control is the most prevalent strategy applied where the primary reactor temperature PID provides a setpoint to a secondary jacket temperature PID for the throttling of hot and cold fluids (Figures 6.1a and 6.1b). The use of cascade control offers considerable performance improvements in terms of reducing the peak and integrated absolute error (IAE) from disturbances. The jacket or coil temperature controller can correct for disturbances to the jacket before they affect the reactor temperature. The jacket or coil temperature loop also isolates process and valve gain nonlinearity from the reactor loop. For a negligible increase in the heat of reaction compared to the heat transfer capability, the process gain for the reactor temperature loop approaches unity cascade control.

> ***Jacket or coil temperature loops prevent the reactor temperature loop from seeing utility system and valve nonlinearities and disturbances.***

For a faster and more uniform measurement the temperature sensor should be in the pipeline about 20 pipe diameters downstream of the outlet of an exchanger or jacket rather than in the equipment for a faster and more uniform measurement of the heat transfer fluid. Pipeline velocity and turbulence reduce the sensor time constant and process noise from the splashing of level, hot and cold spots, and mixed phases. The sensor thermowell tip should be near the centerline of the pipe with a sufficient immersion length to make conductive heat loss negligible.

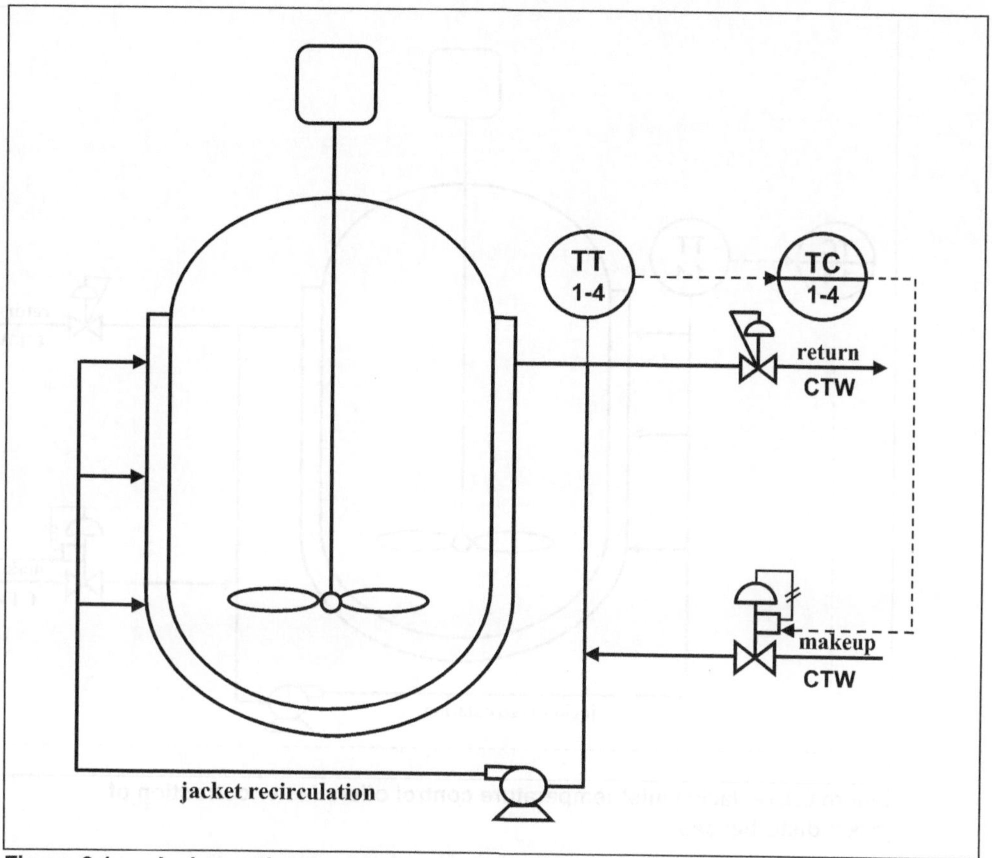

Figure 6.1a – Jacket outlet temperature control offers moderation of mixing nonuniformities, changes in phase, and discontinuities

For highly exothermic fluidized reactors boiler feed water (BFW) is added under level control and the reactor temperature PID output is the jacket outlet steam pressure PID setpoint (Figure 6.2a) or coils are switched in and out of service (Figure 6.2b). The jacket temperature control schemes are suitable for batch as well as continuous operation.

Coils generally offer a faster temperature response than a jacket by a decrease in the volume of the heat transfer fluid and an increase in the velocity. Both of these work to decrease the process dead time, which is the coil volume divided by the utility flow rate. The increase in velocity increases the heat transfer coefficient but this is partially offset by the decrease in surface area. An increase in the product (UA) of the overall heat transfer coefficient (U) and surface area (A) will decrease secondary process lag in the thermal response. The transition in split range operation is faster with a coil than a jacket which is useful for a valid transition between hot and cold utility streams but can be

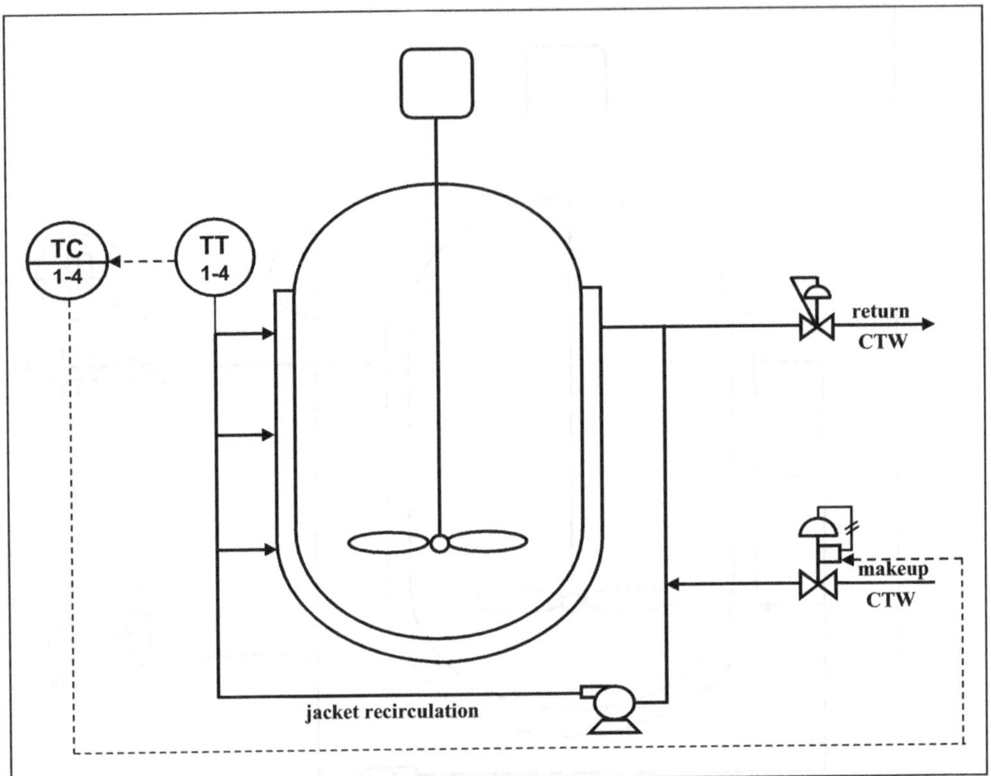

Figure 6.1b – Jacket inlet temperature control offers faster correction of jacket disturbances

problematic for unintended transitions from valve stick-slip and an integrating response in the process or controller.

> ***The more aggressive temperature effect and faster transition between heating and cooling by coils helps deal with the highly exothermic and fast gas reactions but the switching of coils and split range transitions are more disruptive to reactor temperatures.***

Whether the secondary loop uses coil or jacket inlet or outlet temperature is often a matter of tradition for a particular company or process industry. The dynamic response of the cascade control system to reactor disturbances such as feed and reaction rate are the same for coil or jacket inlet and outlet temperature control. Coil or jacket inlet temperature control will correct for changes in cooling or heating utility supply temperature and pressure sooner than the

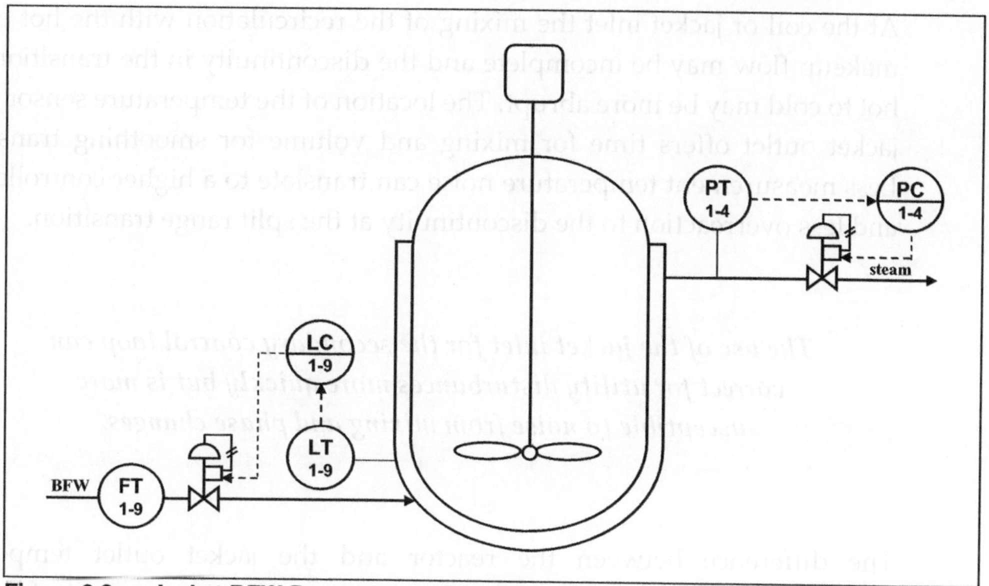

Figure 6.2a – Jacket BFW Pressure and Level Control

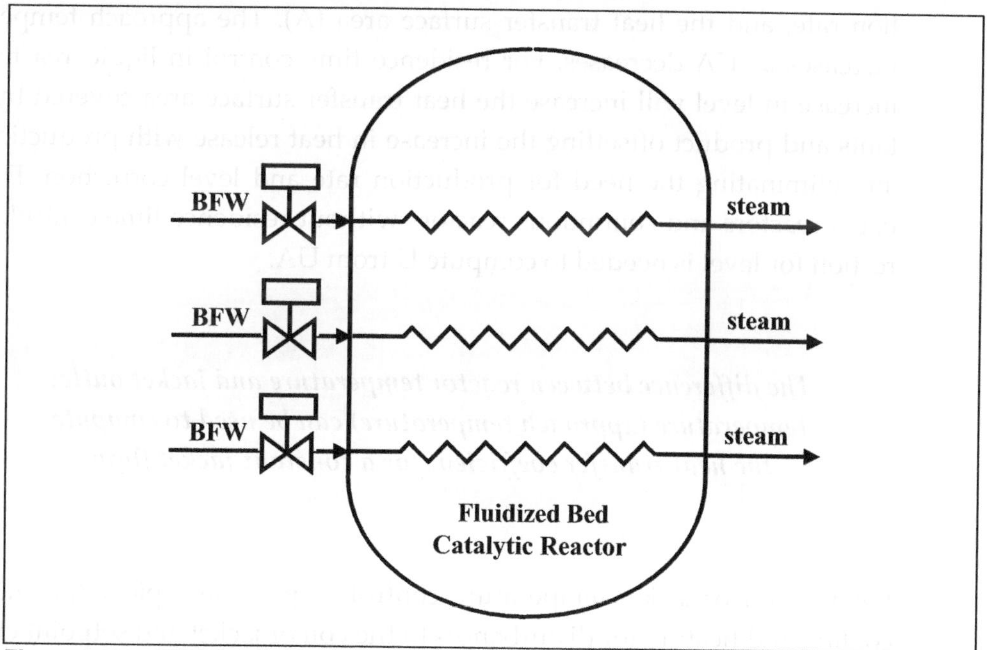

Figure 6.2b – BFW Coil Switching

transportation delay through the coil or jacket. The coil or jacket loop process dead time is also less by the amount of this delay, allowing a faster reset time setting and faster correction of valve nonlinearities.

At the coil or jacket inlet the mixing of the recirculation with the hot or cold makeup flow may be incomplete and the discontinuity in the transition from hot to cold may be more abrupt. The location of the temperature sensor on the jacket outlet offers time for mixing and volume for smoothing transitions. Less measurement temperature noise can translate to a higher controller gain and less overreaction to the discontinuity at the split range transition.

> ***The use of the jacket inlet for the secondary control loop can correct for utility disturbances more quickly but is more susceptible to noise from mixing and phase changes.***

The difference between the reactor and the jacket outlet temperature (approach temperature) can provide an inferential measurement of the heat transfer coefficient (U) for a constant jacket circulation flow, a given production rate, and the heat transfer surface area (A). The approach temperature increases as UA decreases. For residence time control in liquid reactors the increase in level will increase the heat transfer surface area covered by reactants and product offsetting the increase in heat release with production rate and eliminating the need for production rate and level correction. For fed-batch reactors and continuous reactors without residence time control, a correction for level is needed to compute U from UA.

> ***The difference between reactor temperature and jacket outlet temperature (approach temperature) can be used to compute the heat transfer coefficient for a constant jacket flow.***

For the coil or jacket temperature control to provide rapid adjustments of cooling and heating for disturbances to the coil or jacket and setpoint changes from the reactor temperature control, the coil or jacket PID response needs to be fast, which may be achieved by a fast sensor and fast tuning.

A tight fitting sensor bottomed in a thermowell made from a high thermal conductivity metal with a tapered tip near the pipe center line provides a fast measurement. Spring loading can ensure that the sensor sheath is bottomed. The clearance between the sheath outside diameter and the thermowell inside diameter must be minimized and the fluid velocity must be maximized. While grounded thermocouple sensors are a few seconds faster in responding than resistance temperature detector (RTD) sensors, the difference is insignificant compared to the effects of thermowell design and fluid velocity.

Thermowell design details are also important for reactor temperature measurement. The greater sensitivity and lower drift of an RTD is important for jacket as well as reactor temperature measurements. The lower drift reduces maintenance and the higher sensitivity provides faster recognition. The use of RTDs also facilitates more accurate online heat transfer computation for process diagnostics and inferential measurements of reaction rate. Thermocouples are preferred for temperatures above 400 °C where RTD insulation resistance and sensor integrity become problematic.

A high PID gain and a smaller reset time provide fast tuning. The jacket temperature controller has a self-regulating process response with a maximum PID gain that is about half of the open loop time constant divided by the product of the open loop steady state gain and dead time as detailed in Section 1.3 on controller tuning. The minimum reset setting is about 4 times the loop dead time for the jacket temperature PID. The elimination of offset from setpoint in the secondary jacket temperature PID is not as important as an immediate response to setpoint changes from the reactor temperature PID. Consequently, proportional action is more important than integral action and robustness to changes in dynamics is better achieved by increasing the reset time.

RTDs with a tight fit in a tapered thermowell at the pipeline centerline and a PID tuned with more proportional action than integral action provide the fastest secondary loop response.

Most temperature loops tend to develop an oscillation across the split range point during the transition between heating and cooling. The transition creates a discontinuity resulting from the change in utility fluid and the increase in friction and loss of sensitivity in valve seating or sealing. The worst case is often the transition between steam and water coolant due to the huge difference in temperatures and the creation of steam bubbles in water coolant or water droplets in steam. Steam valves tend to have higher seating and sealing friction due to the higher temperatures and pressures. Hot and cold liquids and tempered water systems offer a much smoother transition than between steam and water coolant and reduce nonlinearities for coil and jacket temperature control.

Steam does not provide uniform heating in coils or a jacket. Steam typically collects in the top of the jacket and condensate collects in the bottom. Hot spots can develop around inlets. Thermal shock and steam hammer can damage glass lined vessels. The time required to drive steam completely out of the jacket before cooling water is introduced introduces a considerably delay in the control system. Improper trap design or operation can cause condensate buildup.

The addition of hot water instead of steam directly into the coil or jacket provides a more uniform heat distribution, a dramatically smoother transition between heating and cooling, and a more efficient and maintainable system. For rapid heating, the use of direct steam injection heaters and pressurized water can provide hot water temperatures well above 100°C (Figure 6.3). If the injection heater has hundreds of small orifices, the bubbles are extremely small and are rapidly and quietly mixed into the water. Variable orifice steam injection heaters are not as quiet and the mixing is not as complete. The use of jacket outlet rather than inlet temperature reduces the possibility of bubbles hitting the temperature sensor.

The use of hot water instead of steam for heating eliminates the discontinuities and noise in the transition between heating and cooling. A steam injector can be used to provide a smooth and rapid transition from cold water to varying degrees of hot water.

Heat exchangers in the coil or jacket recirculation system are used to provide a tempered water system. This allows the use of colder or hotter utility streams as inputs. As the inputs are to heat exchangers rather than being direct inputs to coil or jacket, temperature extremes are moderated that would cause heat transfer surface coating or product degradation from localized cold and hot spots thermal shock that could crack a glass lining.

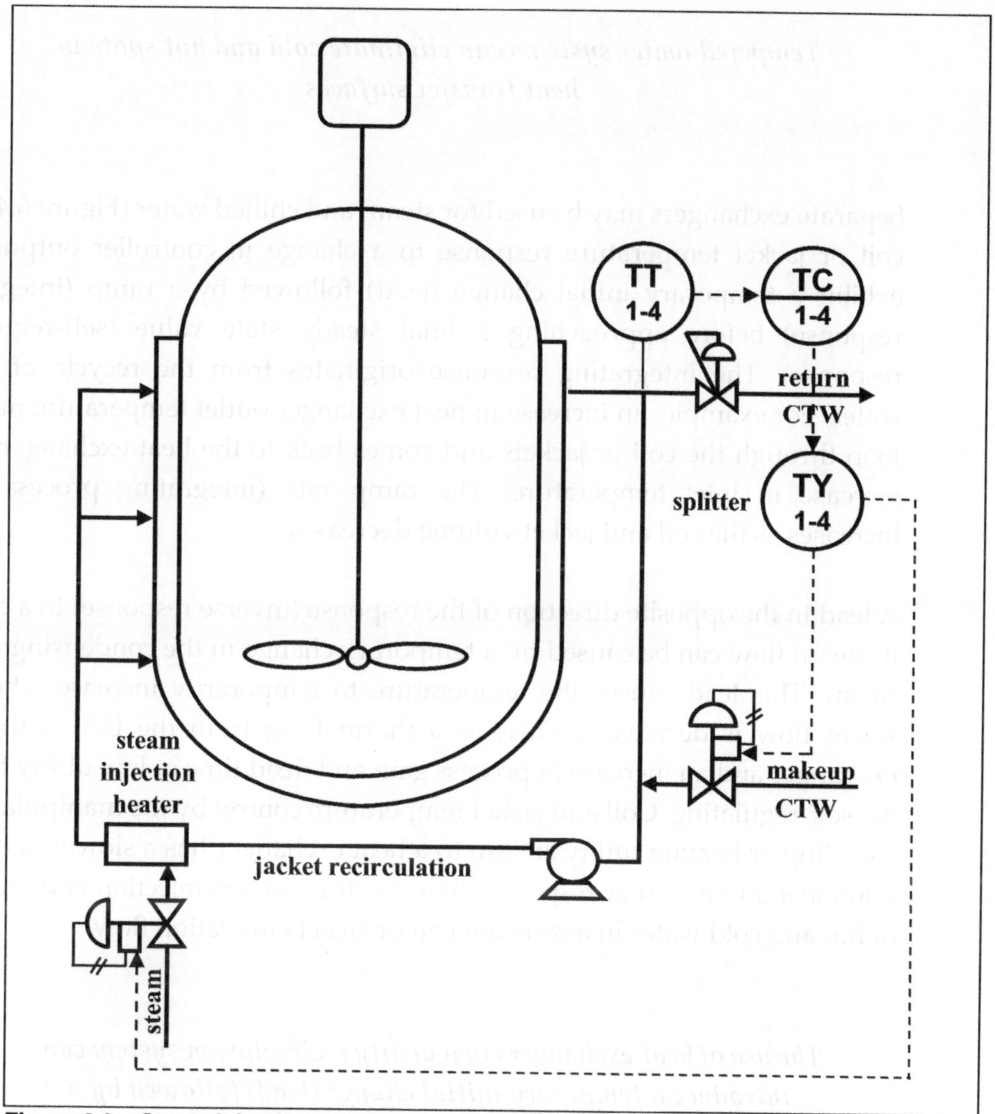

Figure 6.3 – Steam injection heaters to create hotter water offer rapid heating and tight control of coil or jacket temperature with smoother split ranged transitions

As we have seen in Section 5.7, mammalian bioreactors are particularly vulnerable to temperature extremes because of metabolic sensitivity. There is also a greater propensity for localized temperature variations due to less mixing from the reduction in agitation to avoid cell rupture. Mammalian cells, unlike bacterial cells, have cell membranes rather than cell walls, making them more susceptible to damage.

Tempered water systems can eliminate cold and hot spots in heat transfer surfaces.

Separate exchangers may be used for steam and chilled water (Figure 6.4). The coil or jacket temperature response to a change in controller output may exhibit a temporary initial change (lead) followed by a ramp (integrating response) before approaching a final steady state value (self-regulating response). The integrating response originates from the recycle of jacket water. For example, an increase in heat exchanger outlet temperature makes a loop through the coil or jackets and comes back to the heat exchanger as an increase in inlet temperature. The ramp rate (integrating process gain) increases as the coil and jacket volume decreases.

A lead in the opposite direction of the response (inverse response) to a change in steam flow can be caused by a temporary change in the condensing rate of steam. This lead causes the temperature to temporarily increase when the steam flow is decreased. There is a thermal lag from the UA of the heat exchanger and an increase in process gain and dead time at low utility flow in the self-regulating. Coil and jacket temperature control by the manipulation of a cooling or heating utility stream to a heat exchanger has a slower and more nonlinear and irregular response than the direct steam injection and blending of hot and cold water in a constant coil or jacket circulation flow.

The use of heat exchangers in a utility recirculation system can introduce a temporary initial change (lead) followed by a ramp (integrating response) before getting into the heat transfer time constant (lag) of the exchanger volume and heat transfer area.

The coil or jacket exchanger thermal lag can be bypassed by the manipulation of an exchanger bypass flow creating a faster temperature loop that is easier to tune (Figure 6.5a). A VPC can be used to reduce cooling or heating liquid utility flow during low loads and to increase temperature loop turndown (Figure 6.5b). Steam heat exchanger bypass control is not used because of steam blowing into the condensate system at low heating requirements.

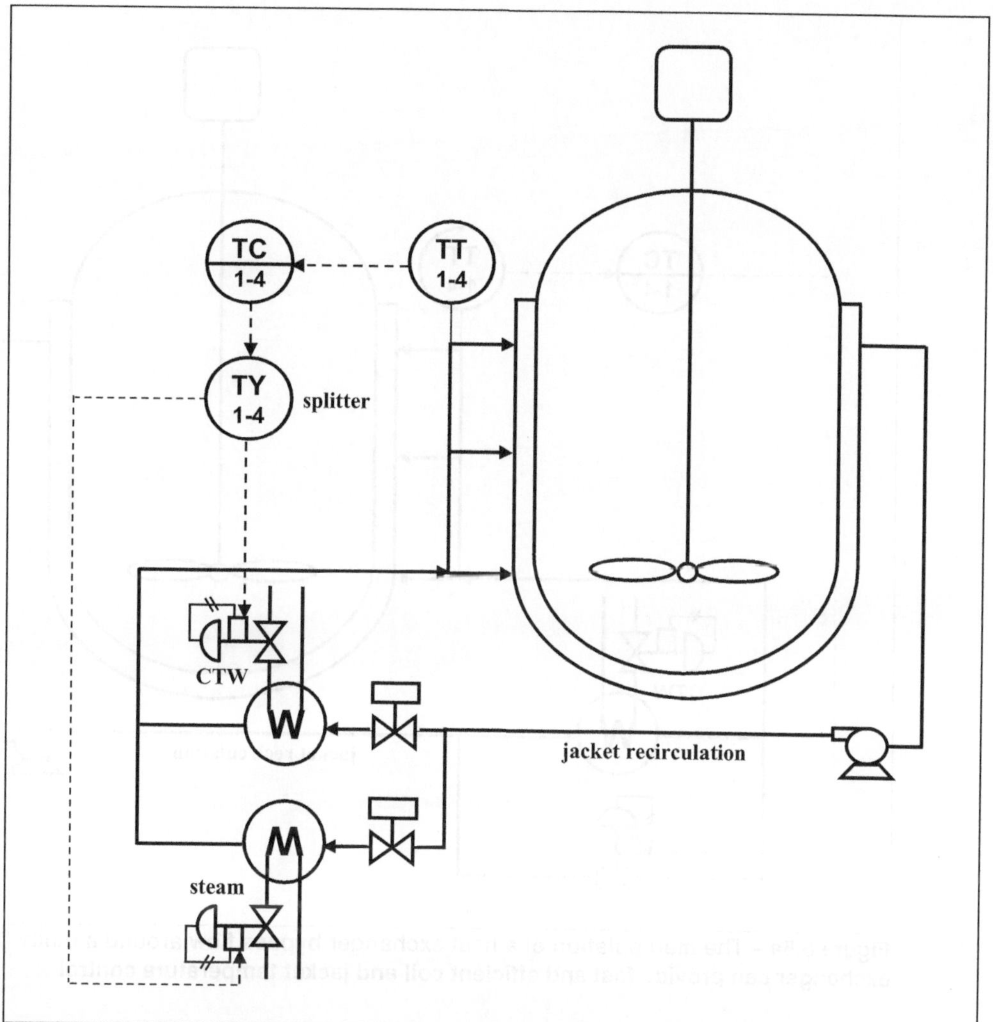

Figure 6.4 – Heat exchangers with split ranged steam and chilled water have less self-regulation, more nonlinearity, and a slower response than direct steam injection and blending of hot and cold water

6-2. Process Recirculation Temperature Control

Heat exchangers in a process recirculation stream are also used for reactor temperature control. The high velocities on the process side in the exchanger increase the heat transfer coefficient and reduce fouling rate. For cascade control the reactor temperature PID output is the setpoint of the exchanger outlet temperature (Figure 6.6a). For the exchanger to do its job the recirculation flow must be high and a wide range of returning process temperature must be tolerated.

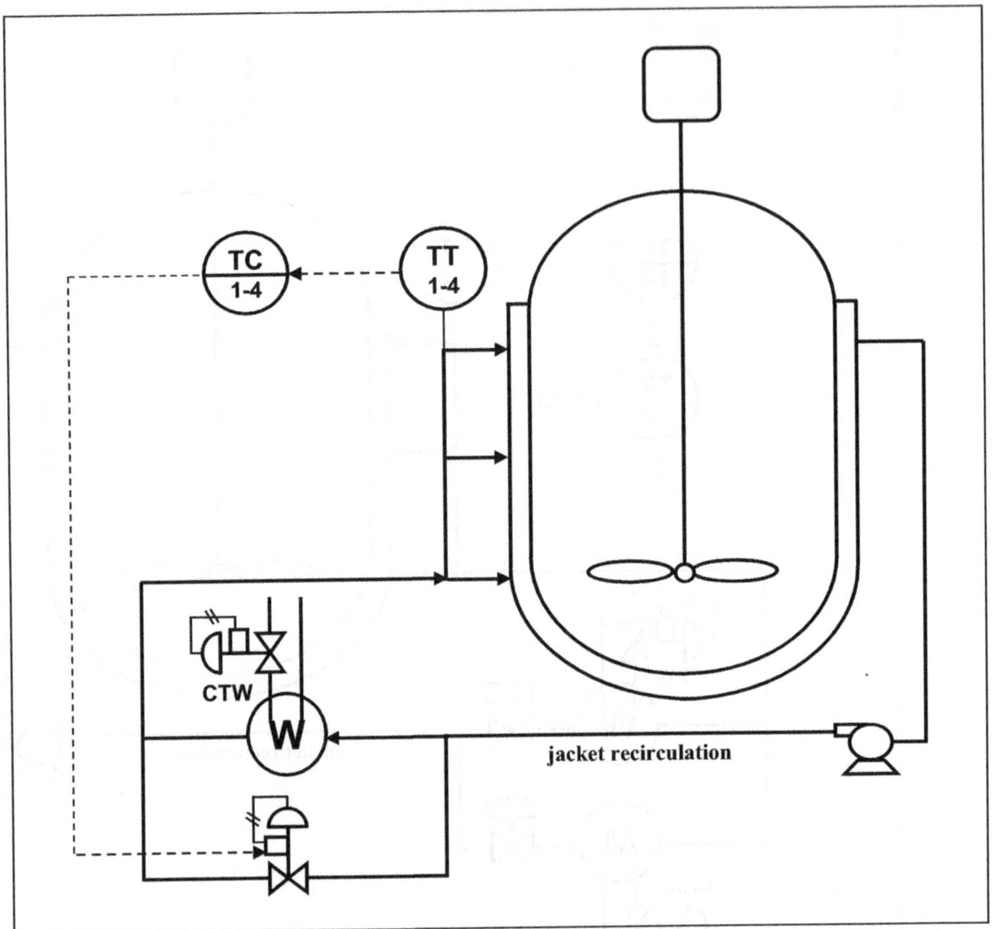

Figure 6.5a – The manipulation of a heat exchanger bypass flow around a utility heat exchanger can provide fast and efficient coil and jacket temperature control

The response of the exchanger outlet temperature has a two stage response. The first stage is a response relatively fast self-regulating response of the exchanger. The second stage response is a protracted integrating response from the recycle effect of the process fluid in the vessel changing temperature. For large vessel volumes, the second stage response ramp rate is slow enough to allow fast exchanger temperature loop tuning based on the initial first stage response. Open loop tests that wait on the exchanger outlet temperature to settle out at a new operating point may be confused by the ramping from the second stage response. Practitioners may inadvertently tune for the second stage response rather than concentrating on having the PID quickly deal with the first stage response.

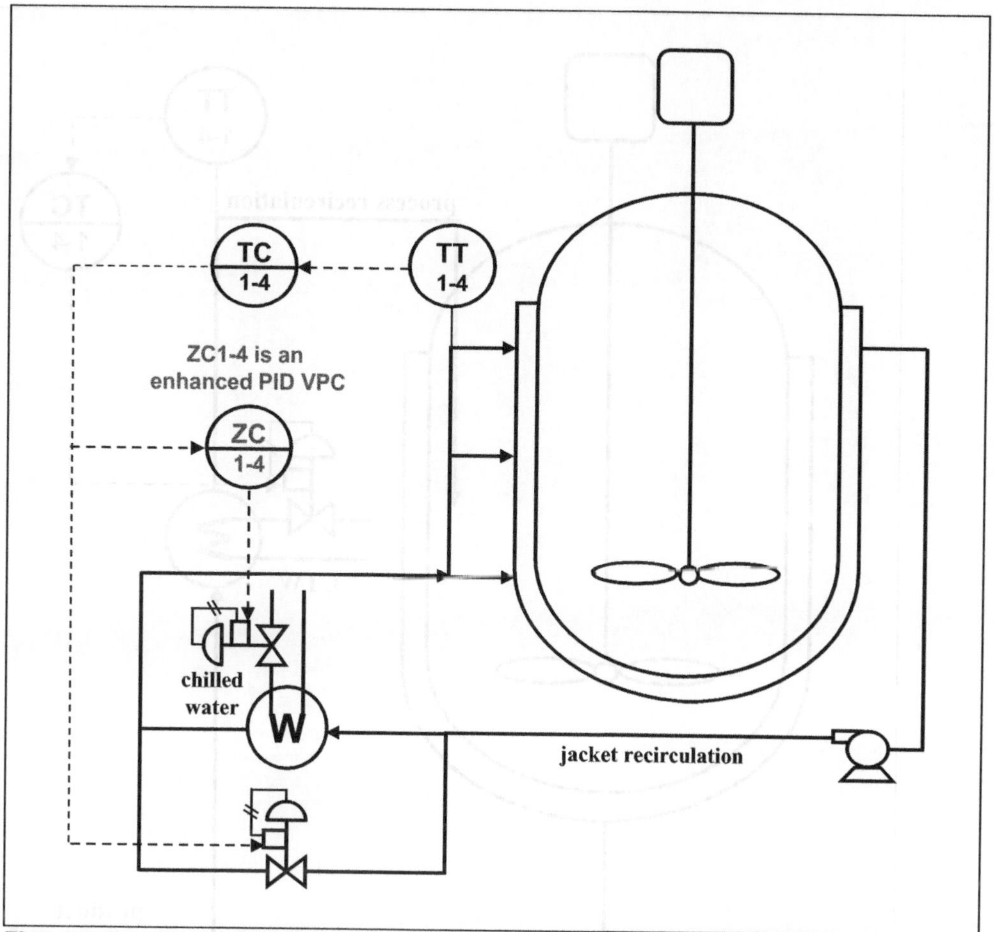

Figure 6.5b – The addition of a valve position controller can reduce utility stream flow and can increase the turndown of coil and jacket temperature control

> ***The throttling of a heat exchanger bypass flow instead of the cooling and heating flow makes the secondary jacket temperature control faster by eliminating the exchanger thermal lag.***

> ***The goal is to make initial self-regulating first stage response much faster than the protracted integrating second stage response of the process recirculation temperature control loop.***

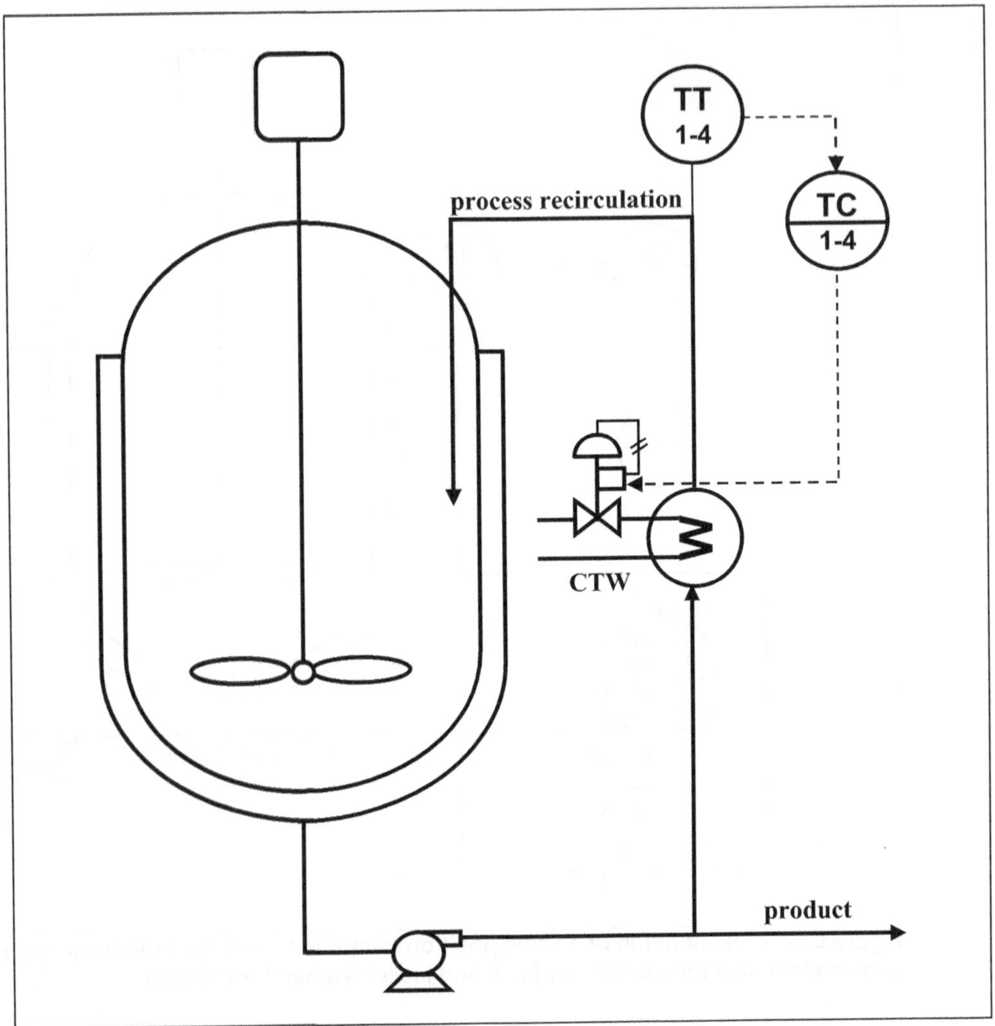

Figure 6.6a – Reactor process recirculation heat exchangers have a two stage temperature response that can confuse open loop tuning tests and users.

The speed of response of control of heat exchanger outlet temperature can be significantly increased by throttling a exchanger bypass flow and keeping the utility flow constant (Figure 6.6b). This mode of operation bypasses the thermal lag of the heat exchanger. The response is as fast as the blending of the streams bypassing and going through the exchanger. Due to the change in heat transfer coefficient with velocity, there is an additional response from the change in heat transfer.

Consider the response of the heat exchanger with cold water. An increase in cooling demand will cause a decrease in bypass flow and an immediate decrease in exchanger outlet temperature. The higher velocity through the exchanger will increase the heat transfer rate to the chilled water, making a further, slower reduction in temperature. If the measurement and the valve are fast enough, the PID can be tuned for a faster rate of response from blending providing further separation between the first stage self-regulating response and the second stage integrating response. Process recirculation heat exchanger bypass control provides a much faster secondary loop for reactor temperature control.

The response of a process recirculation heat exchanger outlet temperature can be significantly faster by throttling a bypass around the exchanger and keeping the utility flow constant.

Faster tuning of the secondary loop can compensate for utility temperature and pressure disturbances before they affect the primary reactor temperature loop, and help to prevent the protracted integrating response from recirculation. A faster secondary loop also makes the primary loop ultimate period faster, enabling more aggressive reactor temperature control and better compensation of process disturbances (e.g., reactant temperature and concentration).

At low cooling or heating requirements, the hot or cold utility valve can be throttled to extend the rangeability of the exchanger. A valve position controller (VPC) can be used to prevent the bypass valve from going wide open by cutting back on the utility flow (Figure 6.6c). A VPC can increase the turndown of the heat exchanger bypass control and prevent the process velocity through the exchanger from going below a minimum, thus reducing the degree of fouling and frosting. This minimum flow also keeps the process side heat transfer coefficient from dropping too low. A low output limit on the VPC can be used to prevent fouling and deterioration of the heat transfer coefficient on the utility side. The VPC should be the enhanced PID to minimize interactions by enabling the use of directional move suppression and eliminating the limit cycles from valve backlash.

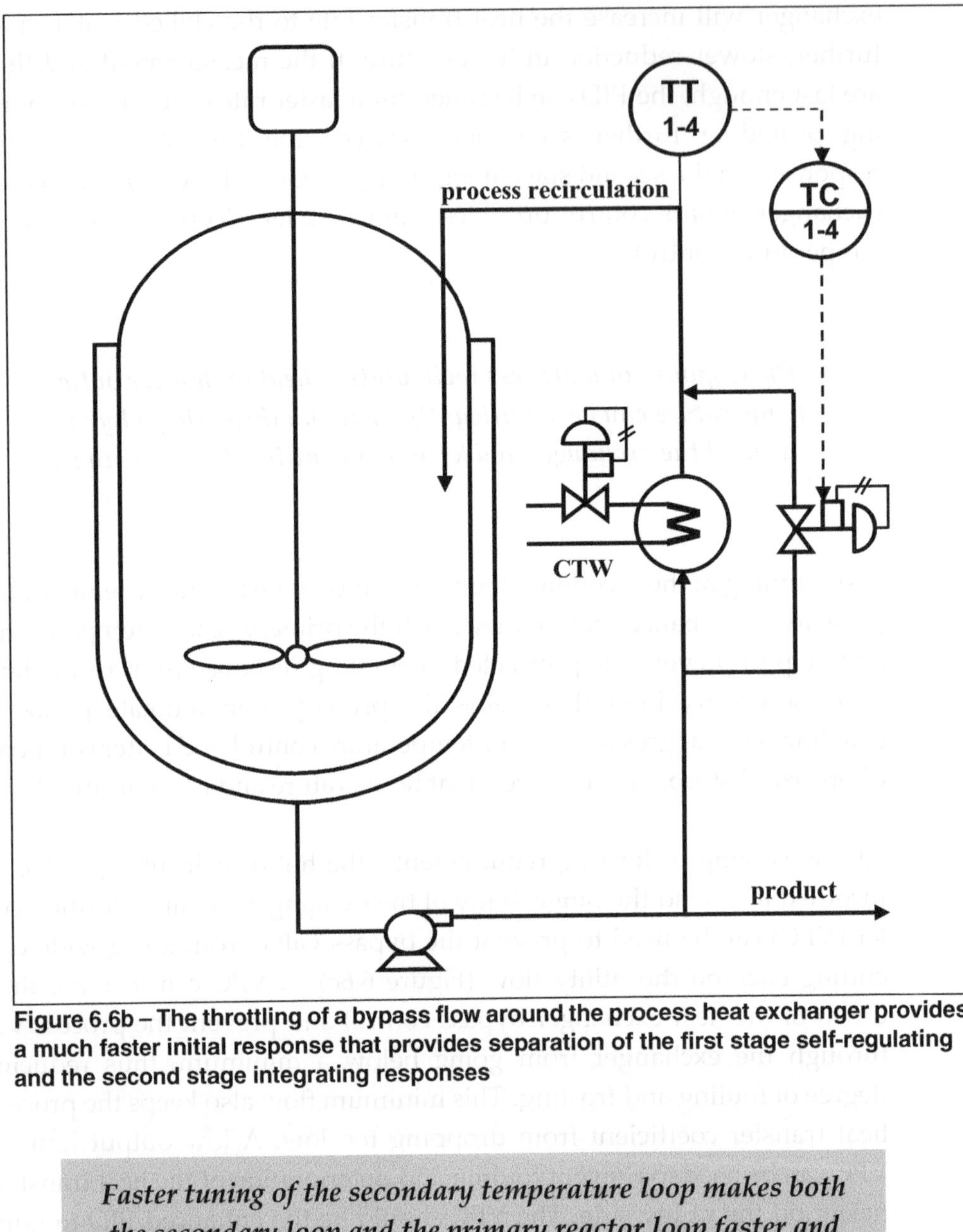

Figure 6.6b – The throttling of a bypass flow around the process heat exchanger provides a much faster initial response that provides separation of the first stage self-regulating and the second stage integrating responses

> ***Faster tuning of the secondary temperature loop makes both the secondary loop and the primary reactor loop faster and eliminates the complications of the protracted integrating response from recirculation. The throttling of a heat exchanger bypass flow enhances this opportunity.***

AVPC can increase the turndown and efficiency of heat exchanger bypass control.

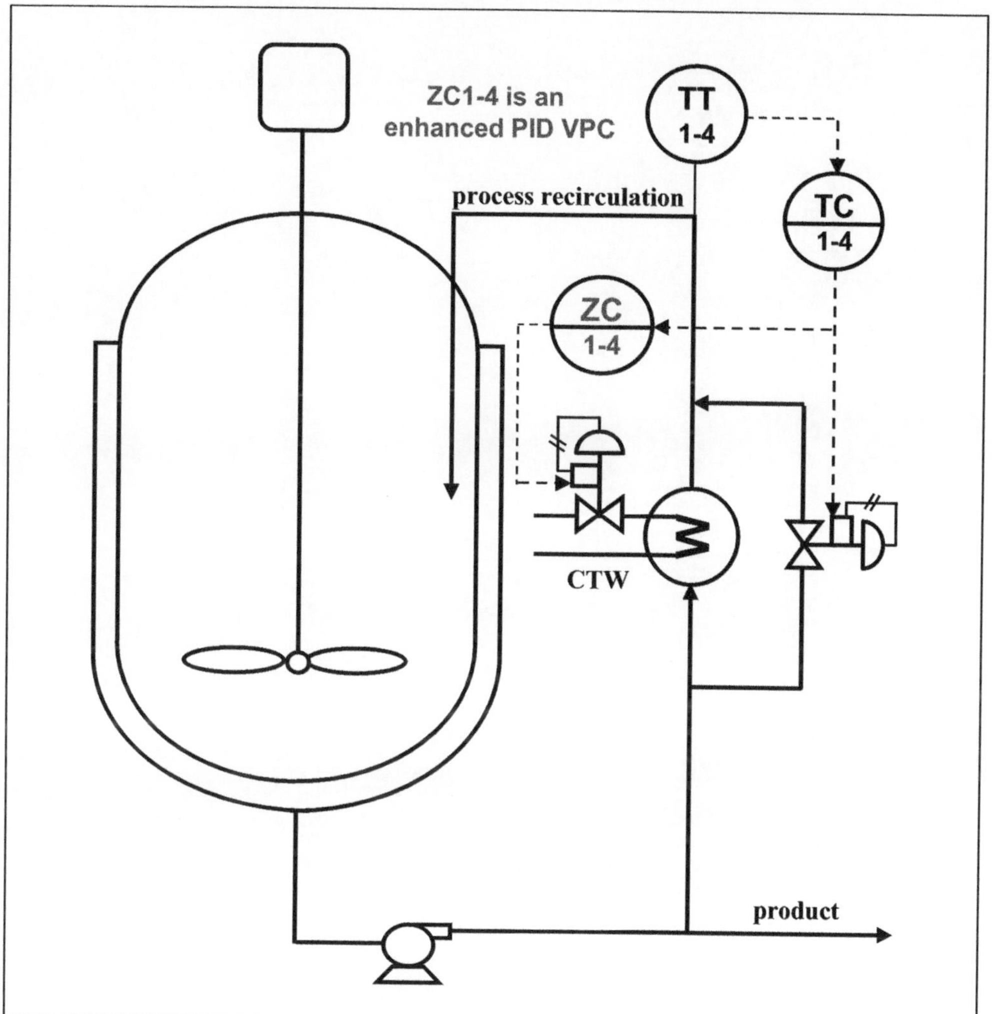

Figure 6.6c – A valve position controller can reduce fouling on the process side and extend the turndown of the temperature loop

Appendix A
Top Ten Concepts in Automation System Performance

Appendix A
Top Ten Concepts in Automation System Performance

Automation system performance not only depends upon tuning but also upon the performance of each component in each control loop. System performance can be maximized by attention to design and implementation details that affect the ability of the controller, final control element(s), equipment, piping, sensor, and transmitter to do their job. Here is a summary of the Top Ten Concepts to guide the automation engineer in this endeavor. We finish with a perspective on the relative important of accuracy and precision.

(1) Delay

- **"Without dead time I would be out of a job"**
- **Fundamentals**
  - A more descriptive name would be **total loop dead time**. The total loop dead time is the amount of time required for the start of a change to completely circle the control loop and end up at the point of origin. For example, an unmeasured disturbance cannot be corrected until the change is seen in the controlled variable and the correction arrives in the process at the same point as the disturbance.
  - Process dead time offers a continuous train of values whereas digital devices and analyzers offer non-continuous data values at discrete intervals, the resulting delays add a phase shift and increase the ultimate period (decrease the natural frequency) like process dead time.
- **Goals**
  - Minimize delay (the loop cannot do anything until it sees and enacts change).
- **Sources**
  - Pure delay from process dead times and discontinuous updates
    - Piping, duct, plug flow reactor, conveyor, extruder, spin-line, and sheet transportation delays (pure process delays set by mechani-

cal design - remaining pure delays set by automation system design)

- Digital device scan, update, reporting, and execution times
- Analyzer sample processing, analysis cycle time, and multiplex time
- Sensor threshold sensitivity
- Signal resolution
- Valve stiction
- Valve backlash
- Variable speed drive deadband
- Wireless trigger level
- Wireless update rate

- Equivalent delay from lags
    - Mixing, column trays, dip tube size and location, heat transfer surfaces, and volumes in series (process lags set by mechanical design - remaining lags set by automation system design)
    - Thermowell lag
    - Electrodes
    - Transmitter damping
    - Signal filters

(2) Speed

- **"Speed kills" - high speed processes and disturbances and with low speed control systems can kill performance**
- **Fundamentals**
    - The rate of change in 4 dead time intervals is most important. By the end of 4 dead times, the control loop should have completed most of its correction. Thus, characterizing self-regulating processes with a large time constant to dead time ratio as near-integrating processes is consistent with performance objectives.
- **Goals**
    - Make control systems faster and make processes and disturbances slower.

- **Sources**
  - Control system
    - PID tuning settings (gain, reset, and rate)
    - Slewing rate of control valves and velocity limits of variable speed drives
  - Disturbances
    - Steps - batch operations, on-off control, manual actions, Safety Instrumented Systems (SIS), startups, and shutdowns
    - Oscillations - limit cycles, interactions, and excessively fast PID tuning
    - Ramps - reset action in PID
  - Process
    - Degree of mixing in volumes due to agitation, boiling, mass transfer, diffusion, and migration

(3) Gains

- **"All is lost if nothing is gained"**
- **Fundamentals**
  - Gain is the change in output for a change in input to any part of the control system. Thus there is a gain for the PID, valve, disturbance, process, and measurement. The open loop gain that is the product of the valve, process, and measurement gain is needed for controller tuning. Knowing the disturbance gain (e.g., the change in manipulated flow per change in disturbance) is important for sizing valves and for feedforward control.
- **Goals**
  - Maximize control system and process gains.
    - Maximize valve sensitivity.
    - Maximize sensor sensitivity.
    - Maximize transmitter gain (minimize calibration span).
    - Maximize controller gain.
    - Maximize process sensitivity (e.g., find best distillation column tray).

- Minimize disturbance gain
  - Minimize size of upstream disturbances by automation (e.g., elimination of oscillations and transfer or absorption of variability by control).
    - Decrease limit cycles.
    - Decrease interactions.
    - Replace operator actions and on-off control with throttling control.
    - Increase absorption of flow variability by surge tank level tuning.
    - Use redundancy to minimize failures (e.g., middle signal selection).
  - Minimize size of process disturbances by enabling process self-regulation (e.g., cancel or moderate disturbances).
    - Get internal reflux control by tight distillate level control.
    - Use weak acid and base reagents to buffer pH systems.

- **Sources**
  - Abnormal operation and failures (e.g., pump, measurement, and valve failure)
  - PID controller gain
  - Inferential measurement location (e.g., sensitivity of temperature change as an inference of composition change in distillation column)
  - Sensor type (e.g., threshold sensitivity of radar versus d/p level measurement and RTD versus thermocouple)
  - Control valve or variable speed drive installed characteristic (gain is slope that includes effect of inherent characteristic and system pressures)
  - Measurement calibration (gain is 100% / span)
  - Process design
  - Backlash and deadband (limit cycle amplitude)
  - Stiction and resolution limits (limit cycle amplitude)
  - Interaction (fighting loops oscillation amplitudes)

- Operator actions
- Safety Instrumented Systems
- Sequences

(4) Resonance

- **"Don't make things worse than they already are"**
- **Fundamentals**
  - Oscillation period close to ultimate period can be amplified by feedback control.
- **Goals**
  - Make disturbance oscillation period slower or control loop faster.
  - For interacting loops with similar priorities, make fast loop faster.
  - For interacting loops with different priorities, make least important loop slower.
- **Sources**
  - Control loops in series with similar dead times (e.g., multiple stage pH control)
  - Control loops in series with similar tuning and valve stiction and backlash
  - Control loops that affect each other with similar dynamics
  - Day to night ambient changes to slow loops (e.g., column temperature control)

(5) Process Attenuation

- **"If you had a blend tank big enough you would not need control"**
- **Fundamentals**
  - Attenuation increases as the volume of the blend tank increases and the ultimate period of the control loop decreases (filter time constant is back mixed volume residence time).
- **Goal**
  - Maximize attenuation by increasing volume and mixing and making loops faster.

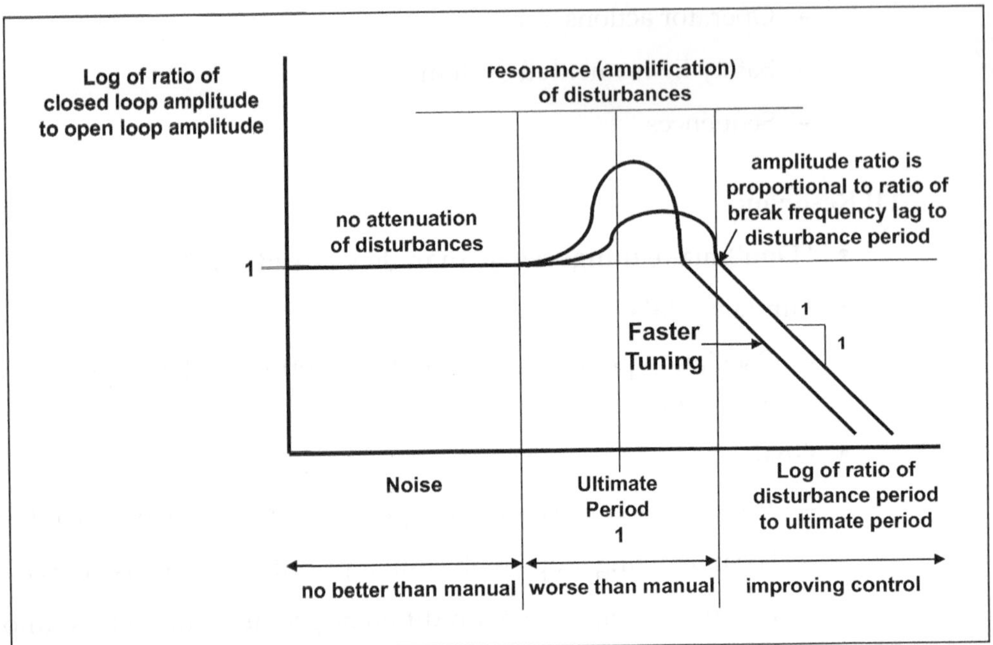

Figure A.1 – An Oscillation with a Period Near the Ultimate Period (Frequency Near the Natural Frequency) is Amplified by the PID from *Resonance*

- **Sources**
  - Mixed volume size and degree of back mixing

$$A_f = \frac{A_o}{\sqrt{1 + (\tau_f * \omega)^2}} \tag{A.1a}$$

The attenuated (filtered) amplitude (A_f) for original amplitude (A_o) and attenuation (filter) time constant (τ_f) that is much greater than the oscillation period (T_o) using the relationship between oscillation frequency and period ($\omega = \frac{2 * \pi}{T_o}$):

$$A_f = \frac{T_o(A_o{}^*)}{2 * \pi * \tau_f} \tag{A.1b}$$

Equation A.1b is also useful for estimating the filtering effect of thermowell lag, electrode lag, transmitter damping, and signal filtering on process oscillation. The equation can also be used to estimate original process amplitude given measurement amplitude when measurement lag is causing an attenuated version of the real process variability.

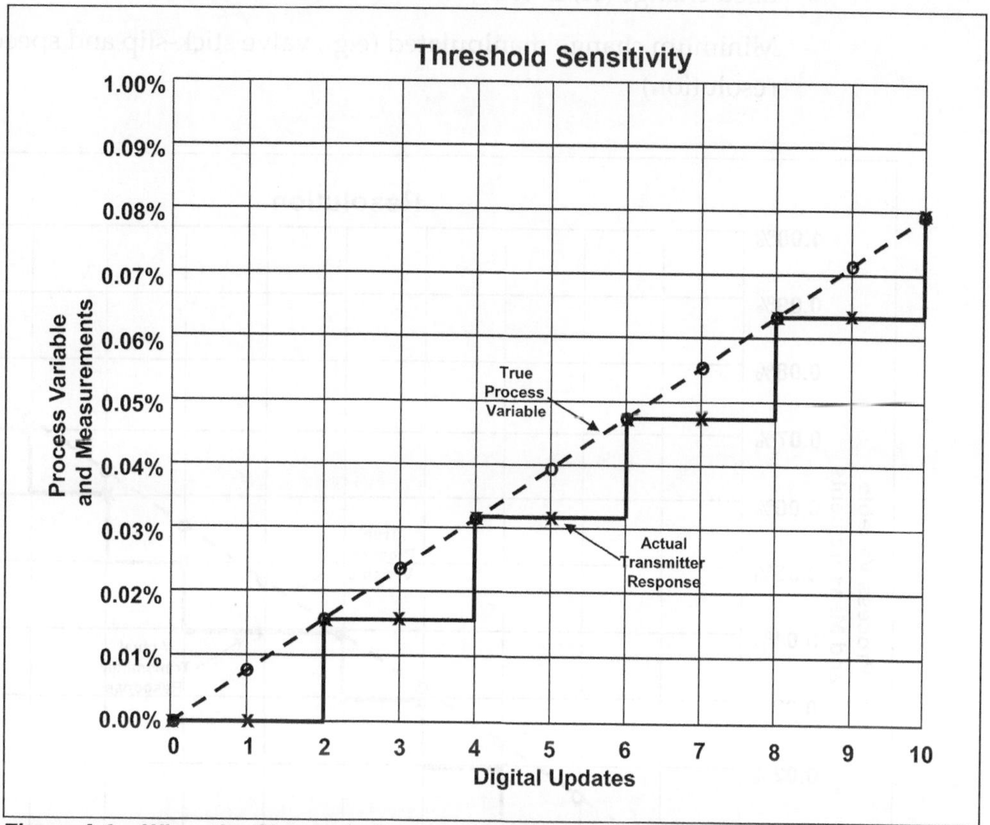

Figure A.2 – When the Change in Process Variable Exceeds the *Threshold Sensitivity Limit*, the Full Change in PV is Seen

(6) Threshold Sensitivity and Resolution

- **"You cannot control what you cannot see"**
- **Fundamentals**
  - Minimum change measured or manipulated - once past threshold sensitivity limit full change is seen or used but resolution limit will quantize the change (stair step where the step size is the resolution limit). Both will cause a limit cycle if there is an integrator in the process or control system.
- **Goal**
  - Improve threshold sensitivity and resolution.

- **Sources**
  - In measurements, minimum change detected and communicated (e.g., sensor threshold and wireless update trigger level) and quantized change (A/D & D/A)
  - Minimum change manipulated (e.g., valve stick-slip and speed resolution)

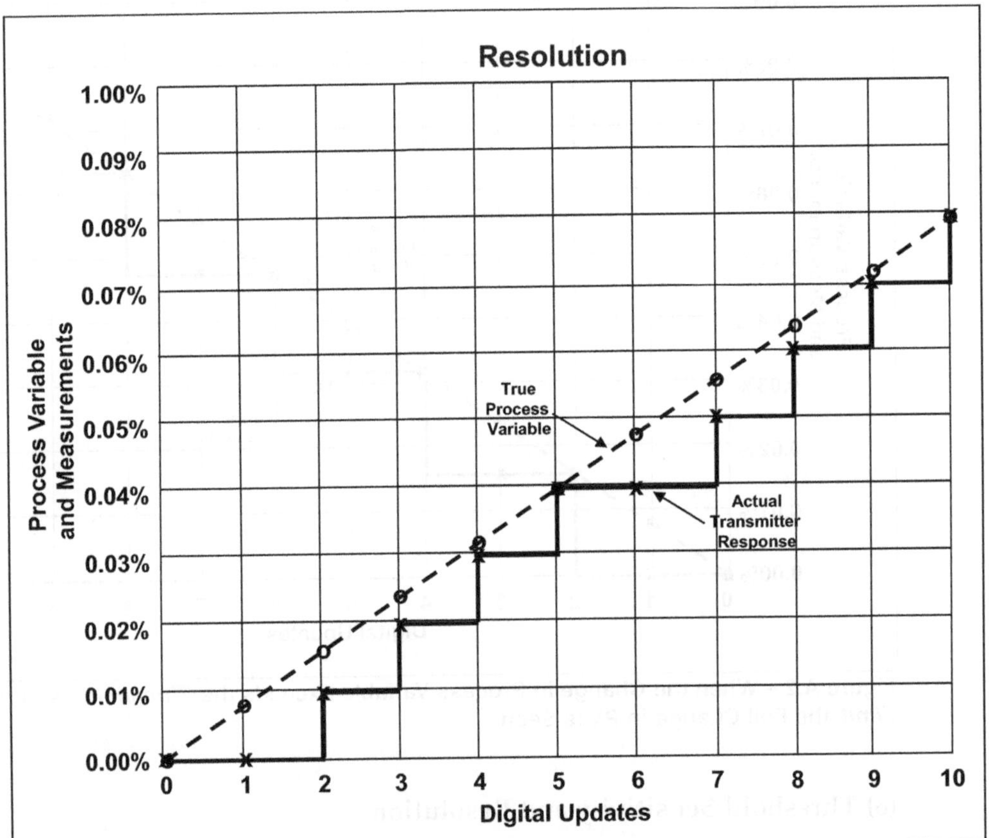

Figure A.3 – When the Change in Process Variable Exceeds the *Resolution Limit*, the Change in PV is Seen as a Step the Size of the Resolution Limit

(7) Backlash Deadband and Hysteresis

- **"No problem man - if you don't ever change direction"**
- **Fundamentals**
  - Hysteresis is the bow in a response curve between full scale traverses in both directions. Normally much smaller and less disruptive than backlash deadband.

- Backlash deadband is minimum change measured or manipulated once the direction is changed - once past backlash-deadband limit you get full change.
- Backlash deadband will cause a limit cycle if there are 2 or more integrators in the process or control system.
- Deadband can be a parameter in a PID or a variable frequency drive (VFD).

- **Goals**
  - Minimize hysteresis in measurements (eliminate mechanical components).
  - Minimize backlash deadband in final control elements.
  - Minimize deadband settings in PID and VFD.
- **Sources**
  - Pneumatic instrument flappers, links, and levers (hopefully these are long gone)
  - Rotary valve and damper links, connections, and shaft windup
  - VFD setup parameter to eliminate hunting and chasing noise

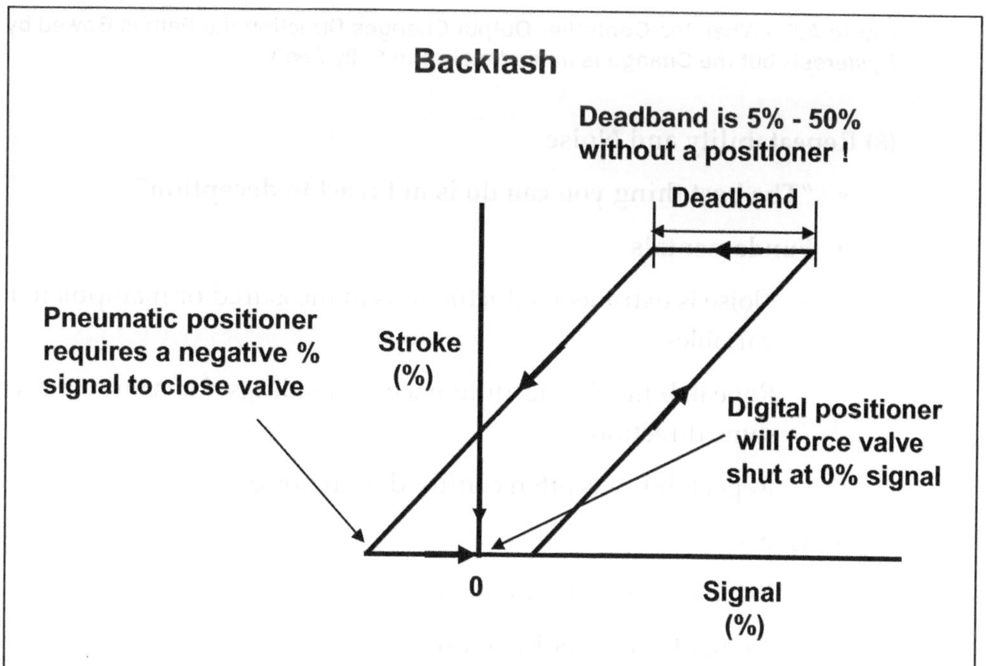

Figure A.4 – When the Controller Output Changes Direction and the Change Exceeds the *Deadband*, the Full Change in Valve Stroke or Drive Speed is Seen

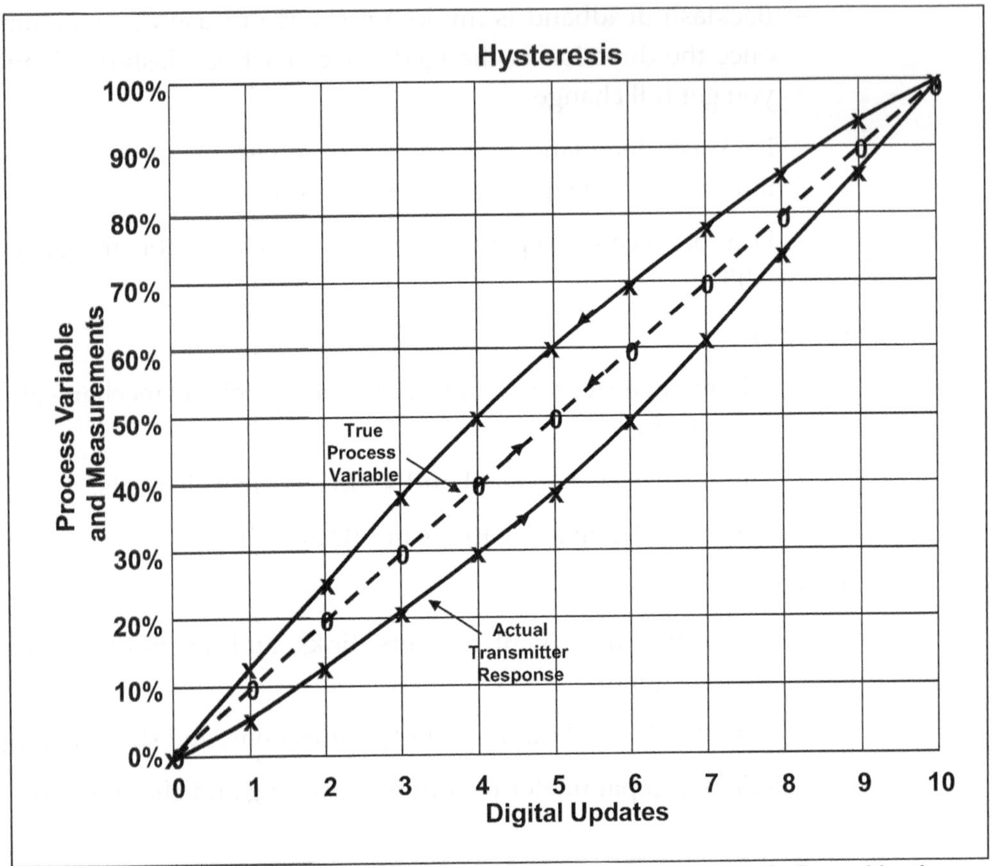

Figure A.5 – When the Controller Output Changes Direction the Path is Bowed by the *Hysteresis* but the Change is Immediately and Fully Seen

(8) Repeatability and Noise

- **"The best thing you can do is not react to deception"**
- **Fundamentals**
  - Noise is extraneous fluctuations in measured or manipulated variables.
  - Repeatability is the difference in readings for same true value in same direction.
  - Repeatability is often confused with noise.
- **Goals**
  - Minimize size and frequency of noise.
  - Do not transfer noise to process.

- **Sources**
  - Noise
    - Bubbles
    - Concentration and temperature non-uniformity from imperfect mixing
    - Electromagnetic interference (EMI)
    - Ground loops
    - Interferences (e.g., sodium ion on pH electrode)
    - Velocity profile non-uniformity
    - Velocity impact on pressure sensors
  - Repeatability
    - Threshold sensitivity, resolution, and extraneous effects

(9) Offset and Drift

- **"There are always an offset and drift; it is matter of their size and consequence."**
- **Fundamentals**
  - Offset is the deviation of the peak in the distribution of actual values from true value.
  - Drift shows up as a slowly changing offset.
- **Goals**
  - Minimize offset and nonlinearity (see Concept 10) by using smart transmitters and sensor matching and smart tuned digital positioners with accurate internal closure member feedback.
- **Sources**
  - Manufacturing tolerances, degradation, de-calibration, and installation effects (process and ambient conditions and installation methods and location)

(10) Nonlinearity

- **"Not a problem if the process is constant, but then again if the process is constant, you do not need a control system."**

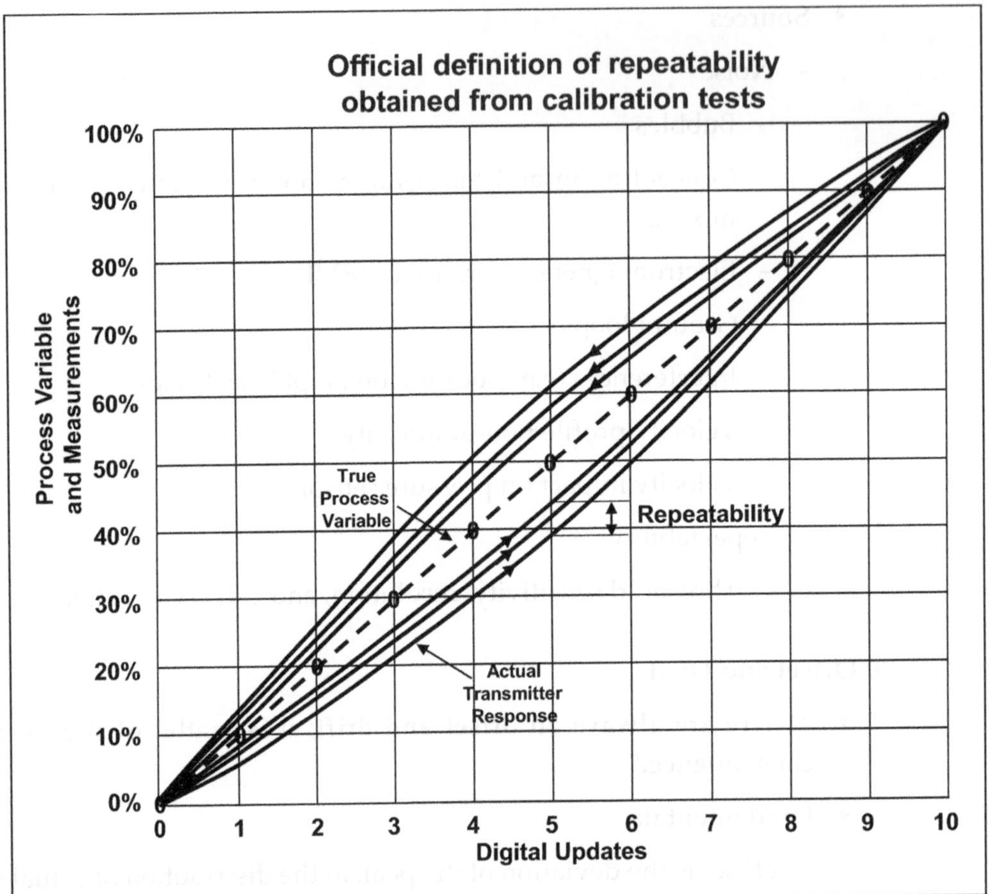

Figure A.6 – The Maximum Separation of the Traverses of the Scale Range in the Same Direction is the *Repeatability*

- **Fundamentals**
  - While normally associated with a process gain that is not constant, in a broader view, a nonlinear system result occurs if a gain, time constant, or delay changes anywhere in the loop. In addition all process control systems are nonlinear to some degree.
- **Goals**
  - Minimize nonlinearity by process and equipment design (e.g., reagents and heat transfer coefficients), smart transmitters and sensor matching, valve selection, signal characterization, and adaptive control.
- **Sources**
  - Control valve and variable speed drive installed characteristics (flat at high flows)

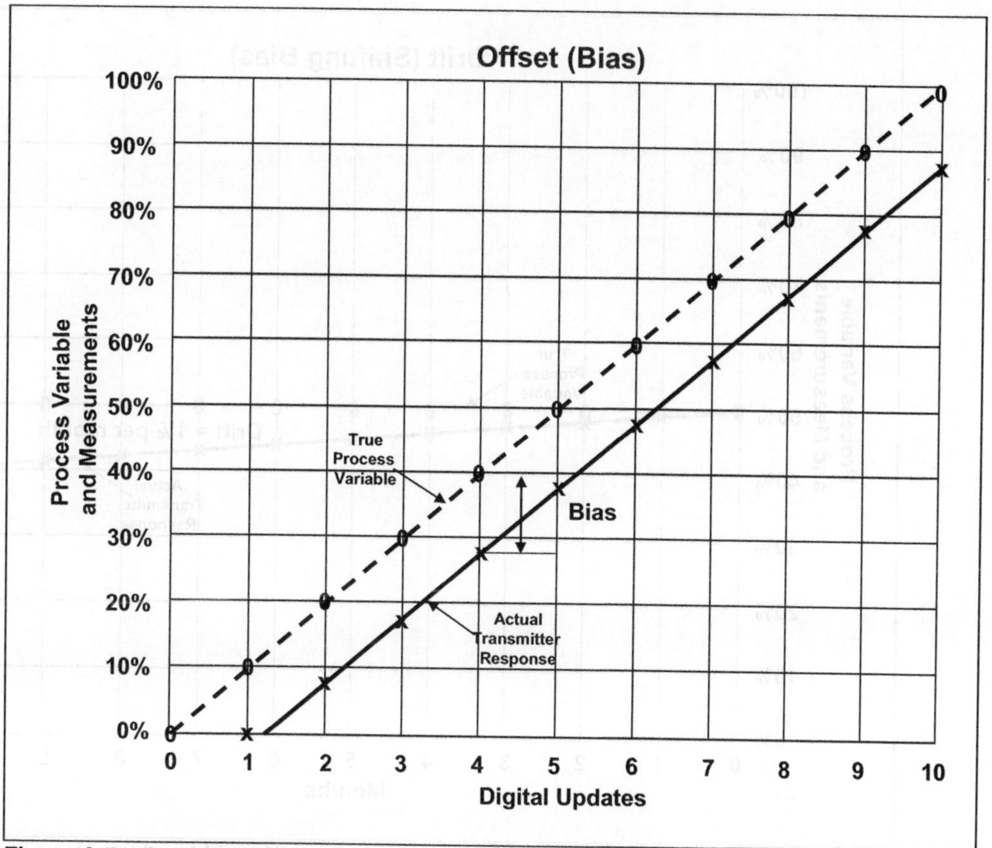

Figure A.7 – An *Offset* (Bias) is a Constant Difference Between the Process Variable and the Measurement in a Complete Traverse of the Scale Range

- Process transportation delays (inversely proportional to flow)
- Digital and analyzer delays (loop delay depends upon when change arrives in discontinuous data value update interval)
- Inferred measurement (conductivity or temperature vs. composition plot is a curve)
- Logarithmic relationship between sensor output and process concentration (glass pH electrode and zirconium oxide oxygen probe)
- Process time constants (proportional to volume and density)

Which is more important, accuracy or precision?

The accuracy of a measurement is determined by the difference between the true value and the measured value. The precision is how much the measured value changes for repeated measurements. In a plot of measurement fre-

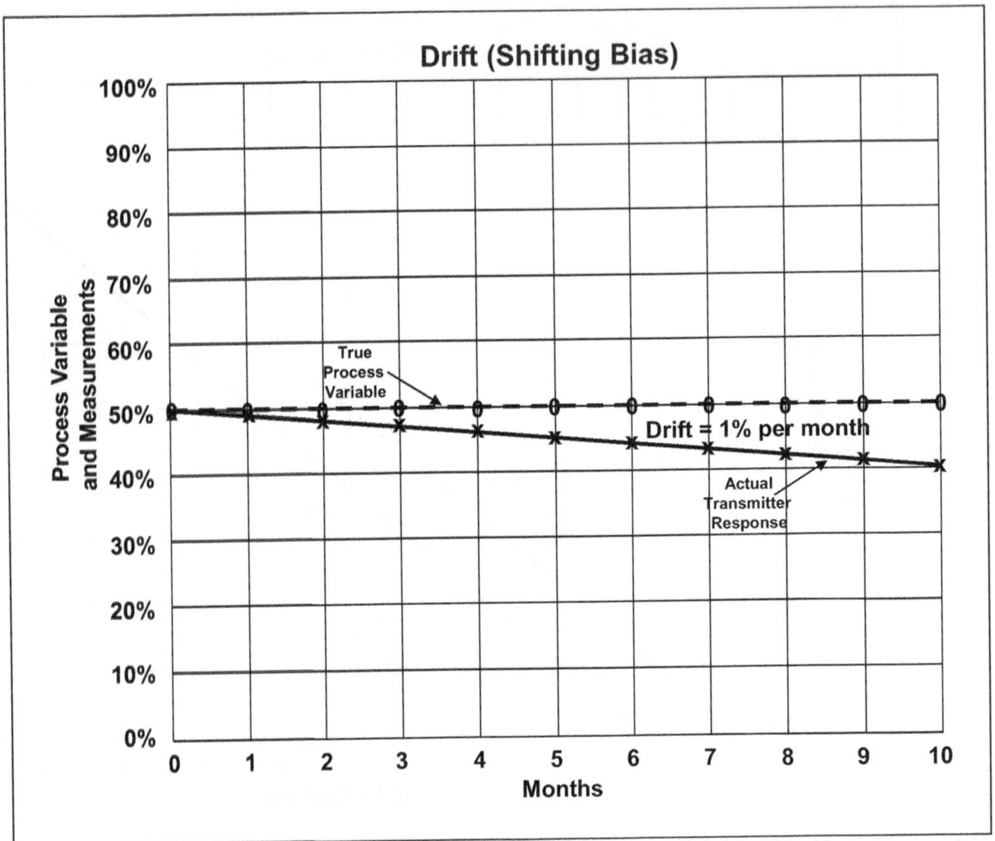

Figure A.8 – A Shifting Difference Between the Process Variable and the Measurement with Time is *Drift*

quency versus measured value for the same true process variable value (Figure A.10), a mean measurement value closer to the true process value is indicative of better accuracy. A tighter distribution of measurement values (small standard deviation), is indicative of better precision.

Inaccuracy can be corrected by changes in setpoint, either manually or by upper level loops, so that the effect is less on performance and more on process calculations and analysis. In contrast, a lack of precision adds an uncertainty for which there are few remedies for the resulting poor loop performance. For example, control valve stroke inaccuracy will introduce errors in system pressure drop and installed flow characteristic calculations but the effect on loop performance will be minimal because the process control loop throttling the valve will adjust the signal to the valve as needed. Similarly, secondary flow measurement inaccuracy will create errors in material balance calculations but the effect on process performance will be minimal

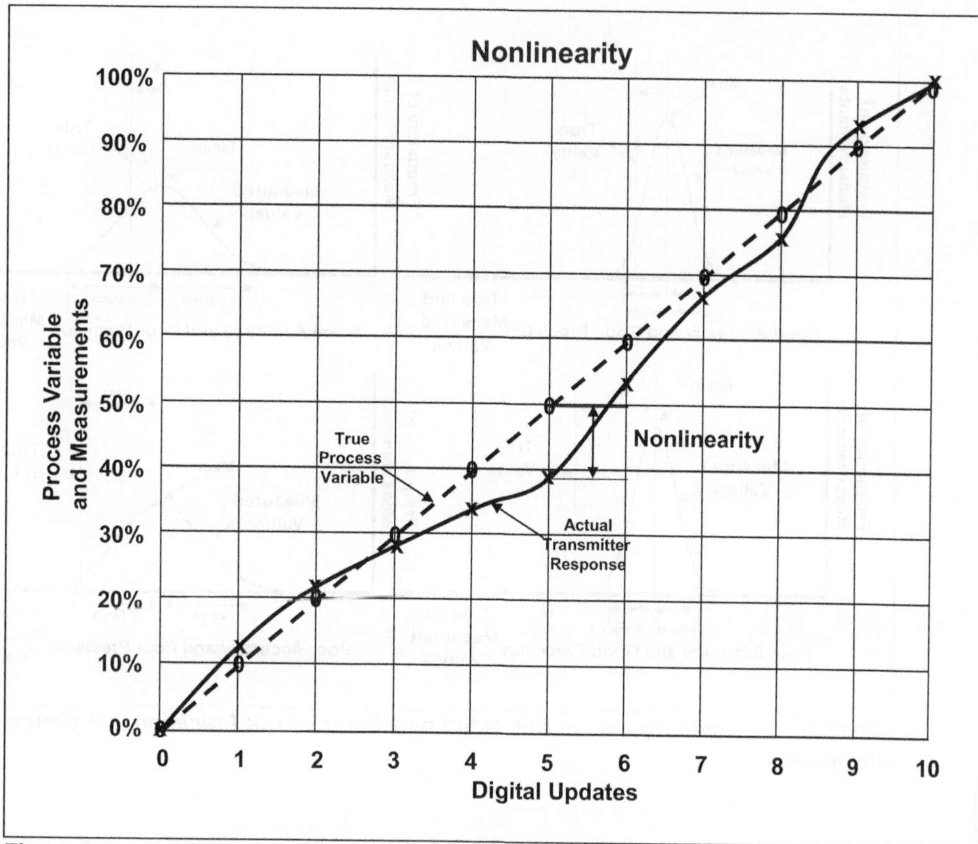

Figure A.9 – A Difference Between the Process Variable Slope and the Measurement Slope for a Traverse of the Scale Range is *Nonlinearity*

because the primary controller in a cascade control system will eliminate the bias by adjusting the secondary controller setpoint.

Consequently, improving factors that affect precision, such as threshold sensitivity, repeatability, and resolution and minimizing backlash, deadband, and noise, are more important than reducing factors that affect accuracy, such as offset, drift, and nonlinearity.

Besides the precision of measurements and final control elements, minimizing dead time and resonance, slowing down disturbances and making them smaller, optimizing process gains, and maximizing process attenuation are important for loop performance. What is left is controller tuning, the primary subject of this book.

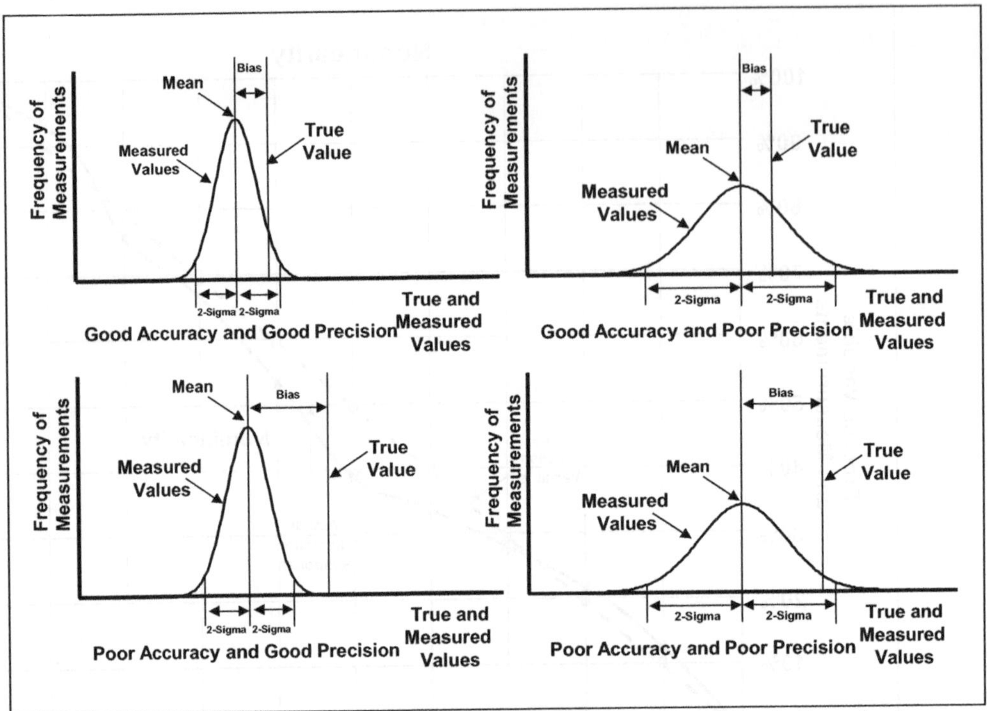

Figure A.10 – *Accuracy* can be Corrected by a Setpoint but *Precision* is Prevalent Uncertainty

Appendix B
Basics of PID Controllers

Appendix B
Basics of PID Controllers

The three PID controller modes each have distinct advantages and disadvantages. Here is a review of the basic functionality of each mode.

Figure B.1 shows the contribution of each mode to the output for a setpoint change. In this example the proportional mode and the derivative mode are acting on the error (the difference between the setpoint and the process variable) rather than just on the process variable. In the time period shown, there is no update from the process measurement so we can see more clearly the function of each mode without any feedback from the process.

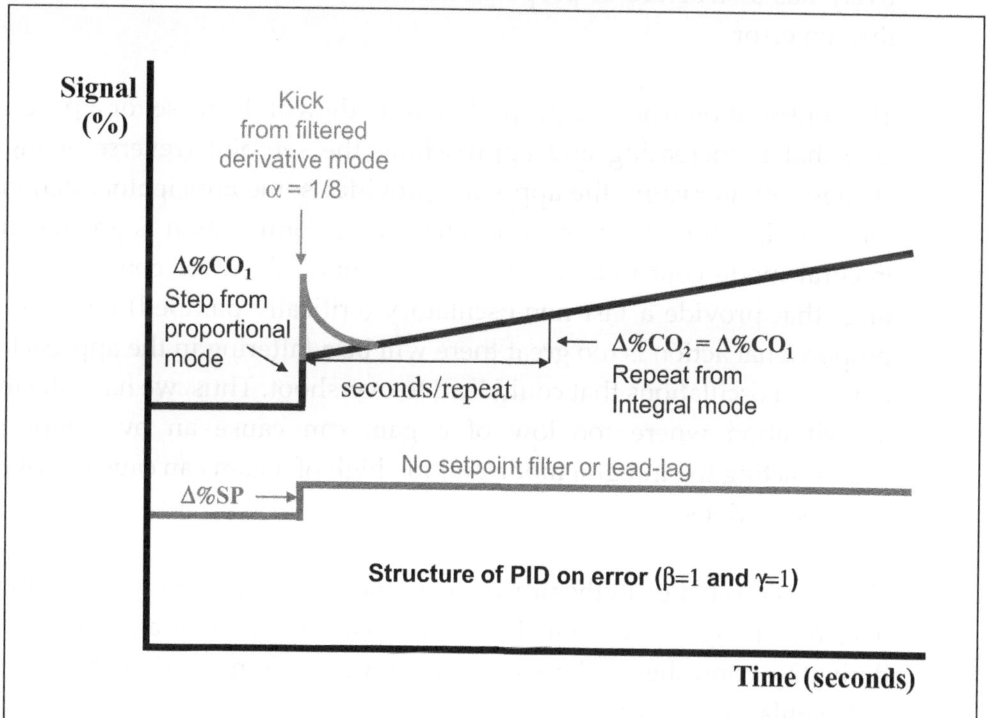

Figure B.1 – Response of Each PID Mode to a Change in Setpoint for PID Action on Error

Proportional Mode Basics

The proportional mode provides an immediate reaction to a change in the process variable. If the PID structure on proportional action on error (β=1), there is an immediate reaction to a setpoint change as well as shown in Figure B.1. The change in percent output is the change in per cent setpoint multiplied by the controller gain. For a step change in setpoint, there is a step in the controller output from the proportional mode. This step will help get the process variable to setpoint sooner and in particular will eliminate the dead time from valve stiction and backlash. However, this step can be too large or disruptive to other loops for some applications. In addition, operators get concerned about rapid changes. The use of a setpoint filter or setpoint rate limit will slow down the change in the PID output and the approach to setpoint, reducing overshoot. The tradeoff is an increase in rise time (the time to reach setpoint). If the setpoint filter time constant is set equal to the reset time, the PID effectively has a structure of proportional action on process variable (PV) rather than on error.

The contribution from the proportional mode will decrease for a process variable that is increasing and approaching the setpoint (reverse acting). This change, counteracting the approach provides some anticipatory action and a sense of direction. If the proportional mode contribution is greater than the integral mode contribution, overshoot is minimized for controller gain settings that provide a fast non-oscillatory (critically damped) response. If the proportional action is too great there will be a faltering in the approach to setpoint and oscillations that could lead to overshoot. Thus, we have the interesting situation where too low of a gain can cause an overshoot by not counteracting the integral mode and too high of a gain can cause an overshoot from oscillations.

If there is no change in the process variable or setpoint, the proportional mode does not change the output. Thus, a proportional only or a proportional plus derivative controller will not cause a limit cycle from stiction or backlash in a self-regulating process.

Note that many analog controllers and one major Distributed Control System (DCS) use proportional band instead of gain for the proportional mode tuning setting. Proportional band is the % change in the process variable (Δ%PV) needed to cause a 100% change in controller output (Δ%CO). A 100% proportional band means that a 100% Δ%PV would cause a 100% Δ%CO, which is a gain of 1. It is critical that users know the units of their controller gain setting and convert accordingly.

Gain = 100% / Proportional Band

Proportional Mode Advantages

- Minimize dead time from stiction and backlash
- Minimize rise time
- Minimize peak error
- Minimize integrated error

Proportional Mode Disadvantages

- Steps in output upset operators
- Steps in output upset other loops
- Amplification of noise

Integral Mode Basics

The integral mode integrates the error with respect to time. For a constant error as shown in Figure B.1, the result is a constant ramp rate. The integral mode will continue to increase the controller output even though the process variable is increasing and approaching setpoint (reverse acting) because the error is negative. Integral action will not change sign until the process variable crosses setpoint. Consequently there is no anticipatory action or sense of direction of change. Integral action that exceeds proportional and derivative action will delay an output in getting off of an output limit and will cause overshoot. Excessive integral action can also cause a runaway reaction by accelerating a temperature that is increasing and approaching a higher setpoint.

If there is no change in the process variable or setpoint, the integral mode will continue to ramp since the error is never exactly zero. Thus, a controller with an active integral mode will cause a limit cycle from stiction in a self-regulating process and from backlash in an integrating process. The "ideadband" (integral deadband) parameter will suspend integral action when the PV is within the specified deadband. The enhanced PID developed for wireless will inherently suspend integral action if there is no PV update. In this case a small threshold sensitivity setting is used to prevent a reaction to noise.

Note that many analog controllers use reset settings in repeats per minute instead of reset time for the integral mode tuning setting. Repeats per minute indicate the number of repeats of the proportional mode contribution in a minute. Today's reset time settings are minutes per repeat or seconds per repeat which gives the time to repeat the proportional mode contribution. The "per repeat" term is often dropped, giving a reset time setting in minutes or seconds.

Seconds per repeat = 60 / repeats per minute

Integral Mode Advantages

- Eliminate offset
- Minimize integrated error
- Smooth movement of output

Integral Mode Disadvantages

- Limit cycles
- Overshoot
- Runaway of open loop unstable reactors

Derivative Mode Basics

The derivative mode provides an output that is proportional to the rate of change of the error with respect to time. For a step change in error, the result is a spike. A built-in derivative filter with a time constant that is fraction α of the rate time, changes the spike to a kick as shown in Figure B.1. Unless the trend chart update time is fast and time scale is short, the kick will look like a spike. When derivative action is used on the PV instead of error, the spike is eliminated since there is rarely a step change in the PV of loops where derivative is used. Derivative is not used in dead time dominant loops because the changes in the PV are abrupt due to the lack of a significant process time constant. If the full response occurs within one dead time, derivative action can do more harm than good. The most effective use of derivative action is the compensation (cancellation) of a secondary lag, particularly in an integrating or runaway process.

Too much derivative action will cause an oscillatory approach to setpoint with a frequency faster than the natural frequency of the loop. These oscillations can persist after the setpoint is reached.

Nearly all derivative tuning settings are given as a rate time in seconds or minutes. The effective rate time setting must never be greater than the effective reset time setting. The effective settings are for an ISA Standard Form. For a PID with Series Form, an interaction factor prevents the effective rate time from becoming larger than the reset time.

The tuning methods developed prior to the 1990s were for the Series Form predominantly in use at the time. These tuning methods may set the rate time equal to the reset time. The use of these settings in an ISA Standard Form as predominantly used today will result in severe oscillations.

The advantages and disadvantages of the derivative mode are similar to those of the proportional mode except the relative advantages are less and the relative disadvantages are greater for the derivative mode.

Derivative Mode Advantages

- Minimize dead time from stiction and backlash
- Minimize rise time
- Minimize peak error
- Minimize integrated error

Derivative Mode Disadvantages

- Kicks in output upset operators
- Kicks in output upset other loops
- Amplification of noise

Split Range Example

I have had several reports from control rooms stating that a PID controller was doing the wrong thing. Consider the temperature control loop where the PV is 48 degrees, the SP is 52 degrees, and the output is split ranged between a water valve and a steam valve with the split range point at 50%. An operator looking at digital values on the operator graphics or the faceplate in Figure B.2 wants the steam valve to be open when in fact the water valve is open. He will ask the for the controller tuning to be corrected to make this happen, which results in a technician or engineer adding more integral action to get the steam valve open. Integral action has no sense of direction and will help meet human expectations and deal with impatience when looking at digital displays.

An operator looking at the trend chart and understanding dead time will realize the temperature was on a trajectory to overshoot the setpoint and any change now will take one dead time to have an effect. The water valve should be open. Improvements in operator displays will continually show relative rate and direction of change and values one dead time into the future.

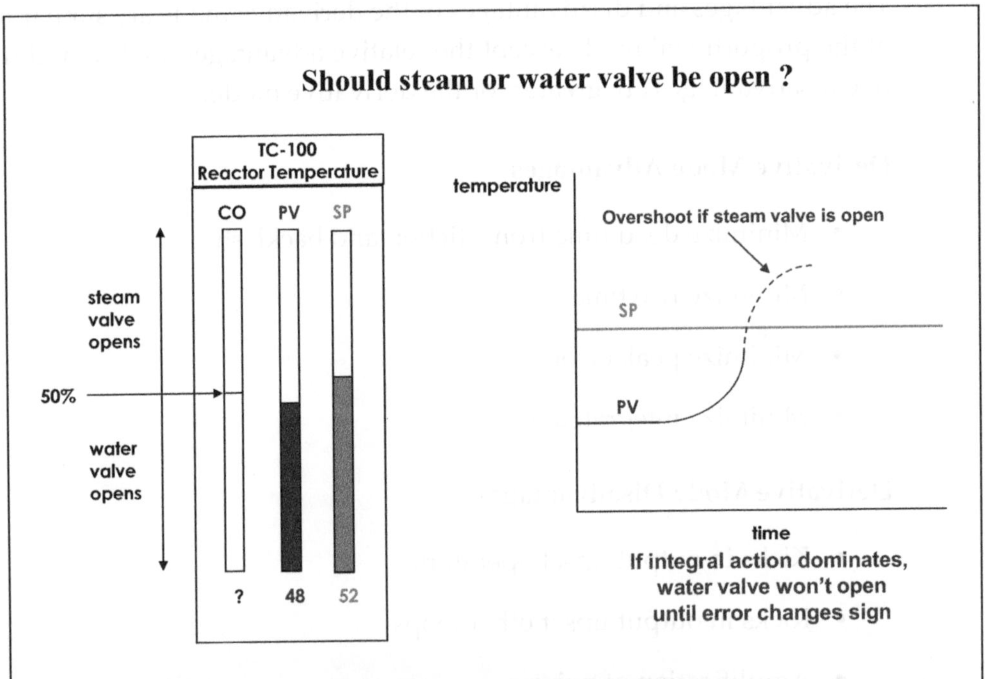

Figure B.2 – An Operator Looking at Digital Values Thinks Steam Valve should be Open Whereas Operator Seeing Trend Trajectory Realizes Water Valve should be Open

Open Loop Gains

The contribution of each mode in the predominant PID Forms used in industry is proportional to the controller gain since this gain is a multiplier in each mode. The controller gain is inversely proportional to a gain that is the change in the percent input for a change in percent output. The process control literature typically specifies this as a process gain without consideration of the contribution of individual components such as installed flow characteristics, feed rate, process variable as a function of feed rate, and measurement span.

To analyze the effects of each contributing component and to account for fundamental differences in the process response, the terms open loop self-regulating process gain or open loop integrating process gain are used instead of simply process gain. The open loop gain is the change in % process variable for a change in % PID controller output for the controller in manual hence the term "open loop." Equation B.1a shows that the open loop gain is the product of the gains for each component in the loop. This breakdown of gains allows one to understand the implications of a change to final control elements, system pressures, production rates, process nonlinearities, and transmitter calibration.

PID Scales

While the user sets and sees the PID process variable and output scales in engineering units, all calculations in the PID algorithm for the contribution of each mode are done in %. Thus, all tuning calculations for controller gain have to include the effect of input and output scale span as shown in Equation B.1a for the open loop gain.

Except for the primary controller in a cascade control system, the PID output scale is 0-100%. When there is a primary controller output scale, it is usually set equal to the secondary controller setpoint scale, effectively cancelling out any effect of the scale ($K_v * K_r = 1$). Consequently, we can focus on the effect of the process variable scale span on the measurement gain to account for the effect of PID scale choice.

We see that the open loop gain is inversely proportional to the process variable scale span. Since the controller gain is inversely proportional to this open loop gain, the controller gain is proportional to the measurement span. Thus, if we decrease transmitter span to improve measurement accuracy, which is often expressed in % of calibration span, we need to proportionally decrease the controller gain for the PID to achieve the same speed of response. If the controller was tuned for maximum disturbance rejection, the controller gain must be decreased to maintain stability. If the PID was tuned with too small of a PID gain, the smaller transmitter span will provide tighter control. Thus, one can understand the confusion about the effects of a transmitter span since changing the span can make a loop do better or worse.

Installed Flow Characteristics

The final control element gain is the change in flow for a change in controller output (Equation 1.1B). This gain is the slope of the installed flow characteristic for the control valve or variable speed drive. For a control valve, the installed flow characteristic is the inherent trim characteristic for a constant pressure drop. For a variable speed drive the installed flow characteristic is linear for zero static head:

$$K_o = K_v * K_r * K_p * K_m \tag{B.1a}$$

$$K_v = \frac{\Delta F_v}{\Delta \%CO} \tag{B.1b}$$

For composition, temperature, and pH loops:

$$K_r = \frac{\Delta R}{\Delta F_v} = \frac{\Delta F_v / F_p}{\Delta F_v} = \frac{1}{F_p} \tag{B.1c}$$

$$K_p = \frac{\Delta PV}{\Delta R} \tag{B.1d}$$

For flow, level, and pressure loops:

$$K_r * K_p = \frac{\Delta PV}{\Delta F_v} \tag{B.1e}$$

$$K_m = \frac{100\%}{(PV_{100\%} - PV_{0\%})} \tag{B.1f}$$

Processes

For composition, temperature, and pH loops in continuous mode, the process gain is the change in slope in a plot of the process variable versus a ratio of the manipulated flow to the process flow. Consequently, the process gain is inversely proportional to the process flow (Equation B.1c).

For flow, level, and pressure loops, the process variable is not a function of this flow ratio. As a result, the process gain simplifies to being just the change in the process variable with the manipulated flow (Equation B.1e).

For integrating processes, such as level, the open loop self-regulating process gain (K_o) is replaced with an open loop integrating process gain (K_i) with inverse time units (1/sec) that results from the process gain term units not cancelling out the flow time units. For mass flow, the level process gain is simply the inverse of the product of the vessel cross sectional area and the liquid density. The process variable is a level in length units (e.g., meters).

Nomenclature for System Gains:

K_m = measurement gain (% / PV e.u.)
K_i = open loop integrating process gain (% / % per sec) (1 / sec)
K_o = open loop self-regulating process gain (% / %) (dimensionless)
K_p = process gain (PV e.u.)
K_r = flow ratio gain often imbedded in process gain (1 / flow e.u.)
K_v = valve or variable speed drive gain (manipulated flow per % output)
$\Delta\%CO$ = change in controller output converted to % of controller output scale (%)
ΔF_v = change in valve or variable speed drive flow (flow e.u.)
F_p = process flow at current production rate (flow e.u.)
$\Delta\%PV$ = change in process variable converted to % of controller input scale (%)
$\%PV_{100\%}$ = process variable at 100% scale value (PV e.u.)
$\%PV_{0\%}$ = process variable at 0% scale value (PV e.u.)
ΔR = change in ratio of manipulated flow to process flow (dimensionless)

Appendix C Control Loop Performance

Appendix C
Control Loop Performance

The following is based on an excerpt from Chapter 14 that the author contributed to *PID Control in the Third Millennium: Lessons Learned and New Approaches*, Editors Ramon Vilanova and Antonio Visioli, Springer 2012.

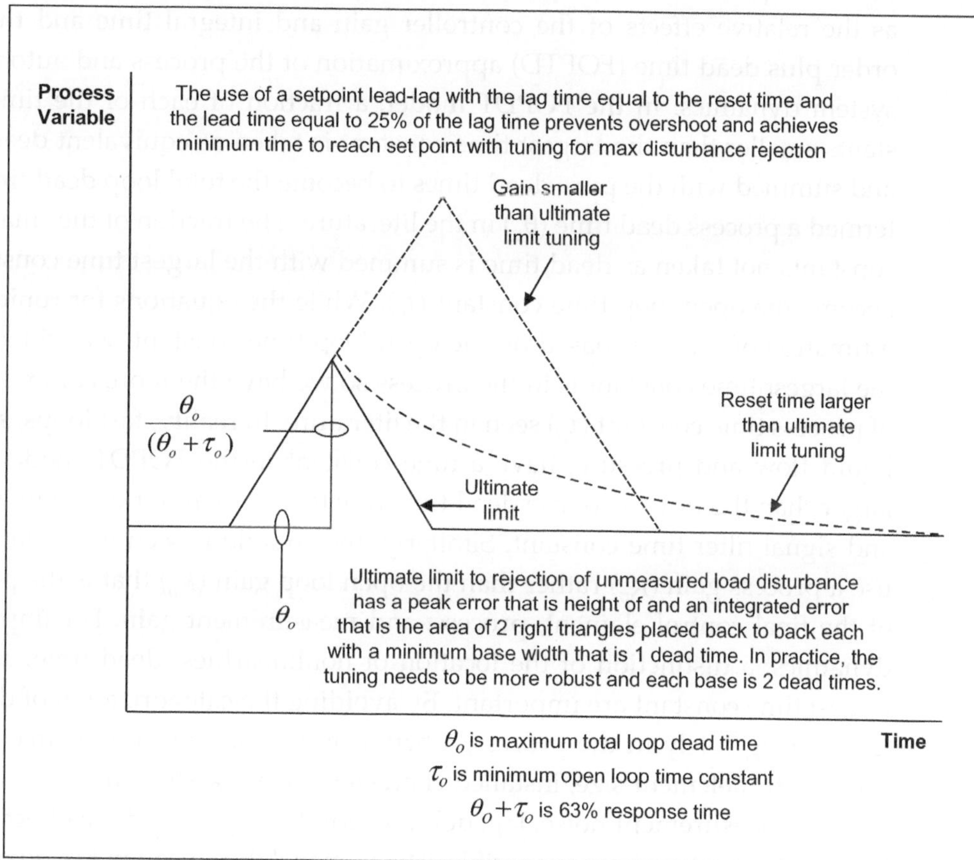

Figure C.1 – Slower than Optimum Tuning Increases the Minimum Peak Error and Integrated Errors that are the Height and Area of an Equilateral Triangle, Respectively

Special algorithms can be designed to deal with measured load disturbances at the process input, setpoint changes, and disturbances at the process output (e.g., noise). Often neglected is the overriding requirement that controllers in industrial applications must be able to deal with unmeasured and unknown load disturbances at the process input. Fortunately, the PID controller excels at this load disturbance rejection. An estimate of the current and best possible load rejection as a function of the process and automation system dynamics and controller tuning provides the information on what can be done to improve plant design and tuning. A simple set of equations can be developed that estimates the integrated error and peak error for a step change in a load disturbance. The value is more in helping guide decisions on improvements rather than predicting actual errors because of the uncertainty of the size and speed of load disturbances and the nonlinear and non-stationary nature of industrial processes. The equations are simple enough to provide key insights as the relative effects of the controller gain and integral time and the first order plus dead time (FOPTD) approximation of the process and automation system dynamics. In the FOPDT model, a fraction of each of the time constants smaller than the largest time constant is taken as equivalent dead time and summed with the pure dead times to become the total loop dead time (θ_o) termed a process dead time (θ_p) in the literature. The fraction of the small time constants not taken as dead time is summed with the largest time constant to become the open loop time constant (τ_o). While the equations for tuning and estimation of errors is based on the open loop time constant, we will assume the largest time constant is in the process so we have the more common term of process time constant (τ_p) seen in the literature. In reality, fast loops, such as liquid flow and pressure, have a time constant in the FOPDT model much larger than the flow response dead time due to a transmitter damping setting and signal filter time constant. Similarly, the equations seen in the literature use a process gain (K_p) rather than the open loop gain (K_o) that is the product of the final control element, process, and measurement gain. For improving dynamics, a distinction of the location of nonlinearities, dead time, and the largest time constant are important. By avoiding the categorization of dynamics as being solely in the process, a better understanding of the effect of the final control element size, installed characteristic, stick-slip, and backlash, the effect of measurement noise, lag, delay, calibration span, and the effect of PID filter and execution time is possible. The nomenclature used in the quantification of these effects is defined at the end of the chapter.

Since a controller cannot compensate for an unmeasured load disturbance before the loop dead time, the peak error (E_x) (maximum error for a disturbance) is the excursion of the first order response to the step disturbance (E_o) based on the open loop time constant for a time duration of the loop dead time (Equation C.1). The open loop error is the final error seen at the PID from an unmeasured load disturbance if the PID was in manual. The terms "open loop" and "closed loop" are used for a response without and with feedback correction, respectively.

$$E_x = [1 - e^{-\theta_o/\tau_o}] * E_o \tag{C.1}$$

If the total loop dead time is much larger than the open loop time constant, then the peak error is basically the open loop error. If the dead time was less than the time constant, then Equation C.1 can be simplified to Equation C.2 eliminating the exponential term. The ultimate limit to the peak error can then be visualized as the ratio of the dead time to 63% response time (dead time plus time constant), which is the altitude of the triangle shown in Figure C.1.

$$E_x = \frac{\theta_o}{(\theta_o + \tau_o)} * E_o \tag{C.2}$$

The minimum integrated error (E_i) can be approximated as the area of two right triangles with the altitude equal to the peak error and the base equal to the dead time as depicted in Figure C.1. The two right triangles form an equilateral triangle. Taking the area of each triangle as ½ the base multiplied by the altitude we obtain Equation C.3 where the integrated error is simply the peak error multiplied by the dead time and consequently proportional to the dead time squared.

$$E_i = \frac{\theta_o^2}{(\theta_o + \tau_o)} * E_o \tag{C.3}$$

Equations C.2 and C.3 are for the minimum possible errors determined by the open loop process and system automation system dynamics. It is not possible to do better than what is permitted by the dynamics of the process and automation system. Thus, Equations C.2 and C.3 provide a benchmark for the ultimate limits to loop performance for unmeasured load disturbances. What is achieved in feedback control depends upon the tuning. In practice, controllers are not tuned aggressively enough to achieve the ultimate limit because the response tends to be too oscillatory especially for large setpoint changes and the controller lacks robustness. A 25% increase in loop dead time or open loop gain or 25% decreases in the open loop time constant can result in oscillations that do not sufficiently decay. Consequently, a more practical benchmark would be right triangles each with a base of two dead times and a time to return to setpoint of 4 dead times.

We can develop the equations that set the practical limit in terms of controller tuning settings from the equations for the ultimate limit based on open loop dynamics. We will also see that we can independently arrive at the same equation for the integrated error from the response of the PI algorithm to a step disturbance.

If we divide through by the dead time term in Equation C.2, we have Equation C.4 where the peak error depends upon the ratio of the open loop time constant to total loop dead time.

$$E_x = \frac{1}{\left(1 + \frac{\tau_o}{\theta_o}\right)} * E_o \tag{C.4}$$

Most tuning methods for maximum disturbance rejection use a controller gain (K_c) that is proportional to the ratio of the open loop time constant to total loop dead time and inversely proportional to the open loop gain (Equation C.5).

$$K_c = \frac{\tau_o}{\theta_o * K_o} \tag{C.5}$$

If we solve for the open loop time constant to total dead time ratio, we see this ratio is simply the product of the controller gain and open loop gain ($K_c * K_o$). If we substitute the product for the ratio in Equation C.3, we have Equation C.6, which is the practical limit to the peak error. In practice an aggressively tuned PID controller with substantial derivative action can achieve this limit with a numerator of 1.1 for a process time constant much greater than the loop dead time. Peter Harriott developed the same form of the equation but with a numerator of 1.5 for the peak error from a proportional only controller tuned for quarter amplitude decaying response:

$$E_x = \frac{1.5}{(1 + K_c * K_o)} * E_o \tag{C.6}$$

For time constant to dead time ratios that are much larger than one, which is the case for pressure and temperature control of vessels and columns, the product of the controller gain and open loop gain is much greater than one leading to the peak error being simply inversely proportional to the product. However due to non-ideal effects the numerator is increased for industrial applications and we end up with Equation C.7 for the peak error:

$$E_x = \frac{2}{K_c * K_o} * E_o \qquad (C.7)$$

Equation C.7 corresponds to a peak error reached in about two dead times. If we approximate the integrated error as the area of two right triangles each with a base equal to two dead times and consider the integral time (T_i) setting as being 4 dead times, we end up with Equation C.8 for the integrated error:

$$E_i = \frac{T_i}{K_c * K_o} * E_o \qquad (C.8)$$

We can derive Equation C.8 from the equation for a PI controller's response to an unmeasured load disturbance. The change in controller output from time *t1* to time *t2* is the sum of the contribution from the proportional mode and the integral mode (Equation C.9a). The module execution time (Δt_x) is added to the reset or integral time (T_i) to show the effect of how the integral mode is implemented in digital controllers. An integral time of zero ends up as a minimum integral time equal to the execution time so there is not a zero in the denominator for the integral mode. For analog controllers, the execution time is effectively zero.

$$\%CO_{t2} - \%CO_{t1} = K_c * (\%E_{t2} - \%E_{t1}) + \left[\frac{K_c}{T_i + \Delta t_x}\right] * \int_{t1}^{t2} \%E_t * \Delta t_x \qquad (C.9a)$$

The errors before the disturbance ($\%E_{t1}$) and after the controller has completely compensated for the disturbance ($\%E_{t2}$) converted to percent of controller input scale are zero ($\%E_{t1} = \%E_{t2} = 0$). Therefore, the long-term effect of the proportional mode, which is first term in Equation C.9a, is zero. Equation C.9a reduces to Equation C.9b:

$$\Delta\%CO = \left[\frac{K_c}{T_i + \Delta t_x}\right] * \int_{t1}^{t2} \%E_t * \Delta t_x \qquad \text{(C.9b)}$$

The integrated error is the integral term in Equation C.9b giving Equation C.9c. For over-damped response the integrated error and the integrated absolute error (IAE) are identical.

$$\%E_i = \int_{t1}^{t2} \%E_t * \Delta t_x \qquad \text{(C.9c)}$$

If we substitute Equation C.9c into Equation C.9b, we have Equation C.9d.

$$\Delta\%CO = \left[\frac{K_c}{T_i + \Delta t_x}\right] * \%E_i \qquad \text{(C.9d)}$$

The change in controller ($\Delta\%CO$) multiplied by the open loop gain (K_o) must equal the open loop error (E_o) for the effect of the disturbance to be eliminated. We can express this requirement as the change in output being equal to the open loop error divided by the open loop gain (Equation C.9e).

$$\Delta\%CO = \frac{E_o}{K_o} \qquad \text{(C.9e)}$$

If we substitute Equation C.9e into Equation C.9d and solve for the integrated error, we end up with Equation C.9f, which is the same as Equation C.8 except for the addition of the execution time interval for the digital implementation of the PI algorithm.

$$\%E_i = \left[\frac{(T_i + \Delta t_x)}{K_o * K_c}\right] * E_o \qquad \text{(C.9f)}$$

Recently, Greg Shinskey added a term to the numerator to include the effect of a signal filter time constant on the integrated error (Equation C.10). In Shinskey's presentation of the equation, the change in controller output rather than the open loop error is used, which eliminates the open loop gain in the denominator. Equation C.10 is applicable regardless of tuning settings. The additional equivalent dead time from the filter time and execution time interval may necessitate a decrease in controller gain and increase in integral time further degrading performance.

$$\%E_i = \left[\frac{(T_i + \Delta t_x + \tau_f)}{K_o * K_c}\right] * E_o \tag{C.10}$$

To summarize, in the process industry, automation system and process dynamics, and in particular the loop dead time, set the ultimate limit to loop performance but controller tuning sets the practical limit for unmeasured disturbances. For example, a loop with a small dead time will perform as badly as a loop with a large dead time if the PID has sluggish tuning. On the other hand, a PID with fast tuning may have an excessive oscillatory response for increases in the loop dead time or process gain. Equation C.6 shows the practical limit to the peak error (E_x) is inversely proportional to 1 plus the product of the PID gain (K_c) and the process gain (K_p). Equation C.9f indicates the integrated error (E_i) is proportional to the ratio of the PID integral time to gain (T_i/K_c). For small filters (τ_f) and PID execution time (Δt_x), the maximum controller gain is decreased, and the minimum integral time is increased based on the increase in loop dead time. The filter and execution time should also be added to the integral time for the integrated error to show the increase in the practical limit (Equation C.10). For a PID tuned for maximum disturbance rejection, Equation C.3 reveals the ultimate limit to the peak error depends upon the ratio of the total loop dead time (θ_o) to the open loop time constant (τ_o). Equation C.8 and C.10 indicates the integrated error depends upon the ratio of the loop dead time squared to open loop time constant. A PID controller tuned for maximum disturbance rejection has a controller gain proportional to the ratio of the largest open loop time constant to loop dead time (τ_o/θ_o), and an integral time proportional to the loop dead time. Note the controller tuning depends upon the largest open loop time constant and not the process time constant. If the largest time constant is in the measurement path, the observed peak error in the measurement predicted by Equation C.2 will be smaller than the actual peak error in the process because of the signal filtering effect of the measurement time constant.

The peak error is important for preventing shutdowns from reaching trip settings of Safety Instrumented Systems (SIS), environmental emissions and process losses from reaching the relief settings of rupture discs and relief valves, off-spec paper sheet and plastic web from exceeding permissible variation in thickness and clarity, compressor shutdowns from crossing surge curve, and recordable incidents by exceeding environmental limits.

The integrated error is a good indicator of the quantity of liquid product off-spec in equipment with back mixing. In these volumes, positive and negative fluctuations in concentration are averaged out unless irreversible reactions are occurring.

An important emerging consideration is the realization that initial open loop response in the FOPDT approximation of a self-regulating process is a ramp seen in the response of an integrating process such as level and batch temperature. The ramp is more persistent in a self-regulating process with a large open loop time constant. The process is termed "near integrating" or "pseudo integrating." An equivalent integrating process gain (K_i) can be approximated as the open time constant divided by the open loop gain (Equation C.11). For processes where the open loop time constant is more than 10 times larger than the dead time, the identification of this near integrator gain in three dead times can reduce the time required for process identification by more than 90% compared to those techniques that go to the 98% response time. Since the PID algorithm works with percent signals, the process variables are converted from engineering units to a percent of PID input scale (%*PV*). The process variable is passed through a dead time block to create an old %*PV* that is subtracted from the new %*PV* to create a Δ%*PV* and then an integrating gain by dividing by the dead time and the change in controller output Δ%*CO*. The maximum of a continuous train of these "near integrating" process gains updated every execution of the PID module can be used for tuning controllers on all types of processes.

$$K_i = \frac{K_o}{\tau_o} \qquad \text{(C.11)}$$

If we substitute the near-integrating gain for the time constant to dead time ratio in Equation C.5, we have Equation C.12. Recently, this method was found to even work on processes where the dead time was greater than the time constant. To provide a smoother response, less setpoint overshoot and more robust settings, the controller gain in both Equation C.5 and C.12 is cut in half.

$$K_c = \frac{1}{\theta_o * K_i} \tag{C.12}$$

The optimum integral time depends upon the type of process. The integral time ranges from about four times the dead time to integrating and "near integrating" processes to ½ of the dead time for severely dead time dominant process ($\theta_o >> \tau_o$). Equation C.13 provides a reasonable curve fit to the required relationship for self-regulating processes. For a dead time less much less than the time constant ($\theta_o << \tau_o$), the ultimate period is about four times the dead time, and the denominator is about one giving an integral time that is about four times the dead time. For a dead time much greater than the time constant ($\theta_o >> \tau_o$), the ultimate period is about two times the dead time, and the denominator is about four, giving an integral time that is ½ of the dead time.

For self-regulating processes:

$$T_i = \frac{T_u}{Min\left(4,\ 10 * \left(\frac{4 * \theta_o}{T_u} - 1\right)^2 + 1\right)} \tag{C.13}$$

For a dead time dominant process, the combination of Equation C.13 for integral time and Equation C.5 for controller gain results in almost an integral-only controller. Since the controller gain is so low, this process is a candidate for setpoint feedforward to reduce the setpoint response rise time.

For an integrating process, the product of the controller gain and integral time must be greater than twice the inverse of the integrating process gain to prevent slowly decaying oscillations from the integral mode dominating the proportional mode. If the user is confident in the knowledge of the integrating process gain, this relationship can be used to find the integral time (Equation C.14a). Since the maximum controller gain allowable on many level and batch temperature loops is greater than 100 and the actual controller gain used is often less than 10, the integral time must be increased to prevent the slow rolling oscillations per Equation C.14a. Consequently, while an integral time of four dead times is possible for an integrating process, in practice an integral time of 40 dead times is more appropriate because the maximum controller gain is beyond the user's comfort level (Equation C.14b).

To prevent large amplitude slowly decaying oscillations for processes with no appreciable deceleration in the PID time frame (4 dead times), the integral action must be greater than ¼ the inverse of the product of the controller gain and integrating process gain.

$$T_i \geq \frac{1}{4 * K_c * K_i} \tag{C.14a}$$

The positive feedback in the runaway processes necessitates an integral time 10 times larger than the integral time for a "near integrating" self-regulating process. The integral time should be 40 dead times or larger for a runaway process (Equation C.14b). Some highly exothermic polymerization batch reactors have gone to proportional plus derivative control to avoid the problem of a user setting too small of an integral time.

For integrating processes with controller gains less than 10 times the maximum controller gain and for runaways (processes with positive feedback) the reset time is increased by a factor of 10:

$$T_i = 40 * \theta_o \tag{C.14b}$$

Too small of a controller gain besides too large of a controller gain can cause a runaway. There is a window of allowable controller gains for positive feedback processes. Any changes in tuning settings particularly for runaway reactions must be closely monitored.

Common metrics for a setpoint response shown in Figure C.2 are rise time (time to reach setpoint), overshoot (maximum error after 1st crossing of setpoint), undershoot (maximum error after 2nd crossing of setpoint) and settling time (time settle out within a specified band around the setpoint). The ultimate limit for rise time is proportional to the loop dead time. The ultimate limit for overshoot and settling time is theoretically zero. The practical limit to rise time is similar to the practical limit for peak error for fast tuning settings but degrades to the relationship for the integrated error for sluggish tuning settings. Fortunately, there are many features that can be used to readily help achieve the ultimate limit to the rise time. The practical limits for overshoot and settling time depend upon a balance between the contributions from the integral and proportional modes. The controller gain for maximum disturbance rejection can be used to minimize rise time, and the integral time can be increased to minimize overshoot and settling time.

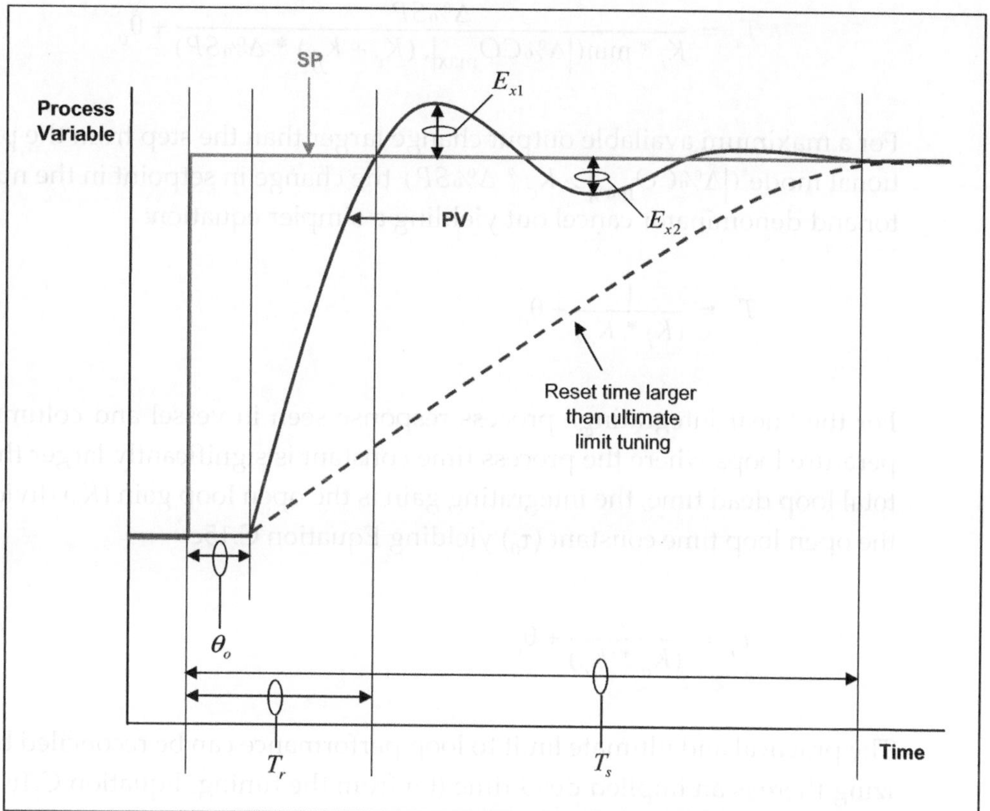

Figure C.2 Aggressive Settings for Load Rejection Minimizes Rise Time (T_r) but Increases Overshoot (E_{x1}), Undershoot (E_{x2}), and Settling Time (T_s) in the Setpoint Response

The minimum rise time (T_r) can be approximated as the change in percent setpoint ($\Delta\%SP$) divided by the maximum rate of change of the process variable. For an integrating or "near integrating" process, the maximum PV ramp rate is the integrating process (K_i) gain multiplied by the change in controller output as detailed in the denominator of Equation C.15a. If the step change in controller output from the proportional mode for a structure of proportional action on error is less than the maximum available output change (difference between current output and output limit), Equation C.15a simplifies to Equation C.15b for feedback control. The output change must be corrected for methods used to make the setpoint response faster. For setpoint feedforward, the step change in output is a combination of the feedforward and feedback action. For smart bang-bang logic, the step output change is the maximum available output change.

$$T_r = \frac{\Delta\%SP}{K_i * \min(|\Delta\%CO_{max}|, (K_c + K_{ff}) * \Delta\%SP)} + \theta_o \tag{C.15a}$$

For a maximum available output change larger than the step from the proportional mode ($|\Delta\%CO_{max}| > K_c * \Delta\%SP$) the change in setpoint in the numerator and denominator cancel out yielding a simpler equation:

$$T_r = \frac{1}{(K_i * K_c)} + \theta_o \tag{C.15b}$$

For the "near integrating" process response seen in vessel and column temperature loops where the process time constant is significantly larger than the total loop dead time, the integrating gain is the open loop gain (K_o) divided by the open loop time constant (τ_o) yielding Equation C.15c.

$$T_r = \frac{\tau_o}{(K_o * K_c)} + \theta_o \tag{C.15c}$$

The practical and ultimate limit to loop performance can be reconciled by realizing there is an implied dead time (θ_i) from the tuning. Equation C.16 shows the implied dead time that can be approximated as the original dead time (θ_o) multiplied by a factor that is 0.5 plus lambda (λ). Lambda is the closed loop time constant for a setpoint change. For a PID tuned for maximum disturbance rejection, lambda is set equal to the original dead time. The implied dead time is then equal to the original dead time.

$$\theta_i = 0.5 * (\lambda + \theta_o) \tag{C.16}$$

The peak and integrated errors for unmeasured step disturbances represents the worst case. Step disturbances originate from manual actions, safety, switches, and sequential operations. If discrete actions (e.g., the opening and closing of on-off valves and the starting and stopping of pumps) are replaced by control loops with modulated final control elements (throttling valves and variable speed drives) or are attenuated by intervening volumes, the step disturbances are smoothed. The attenuated load disturbance has a time constant (τ_L) that is the residence time of the volume or closed loop time constant of the upstream control loop. To include the effect of a load time constant, the process excursion in the first dead time, which is the key time for determining minimum peak error, can be computed by Equation C.17. The open loop error (E_o) in the equations for peak and integrated error can be replaced with a lagged load disturbance (E_L) that is the open loop error multiplied by the exponential response of the disturbance in one dead time. The effect is mitigated by a reset time that is slow relative to the disturbance time constant.

$$E_L = (1 - e^{-\theta_o/\tau_L}) * E_o \qquad \text{(C.17)}$$

PID controllers tuned too fast can introduce process variability from an oscillatory response. PID controllers tuned too slowly can make a loop with good dynamics perform as badly as a loop with poor dynamics. In other words, money invested to reduce process dead time or to get faster measurements and valves is wasted unless the PID controller tuning is commensurate with the speed of the process so that the practical limit approaches the ultimate limit to loop performance.

Nomenclature

$\%CO_{t1}$ = controller output at time $t1$ before correction for load disturbance (%)

$\%CO_{t2}$ = controller output at time $t2$ before correction for load disturbance (%)

$\Delta\%CO$ = change in controller output converted to % of controller output scale (%)

$\Delta\%CO_{max}$ = maximum available change in controller output (%)

$\Delta\%PV$ = change in process variable converted to % of controller input scale (%)

$\Delta\%PV_{max}$ = maximum change in process variable in one dead time interval (%)

$\Delta\%SP$ = change setpoint converted to % of controller input scale (%)

$\%E_i$ = integrated error from load disturbance (% * sec)

E_L = open loop error corrected for load disturbance time constant (%)

E_o = open loop error (load disturbance error with controller in manual) (e.u.)

$\%E_t$ = error at time t from load disturbance converted to % of controller input scale (%)

$\%E_{t1}$ = error at time $t1$ from load disturbance converted to % of controller input scale (%)

$\%E_{t2}$ = error at time $t2$ from load disturbance converted to % of controller input scale (%)
E_x = peak error from load disturbance (e.u.)
K_c = controller gain (dimensionless)
K_i = integrating process gain (% per sec per % = 1/sec)
K_o = open loop gain (%/%) (dimensionless)
K_p = process gain (e.u./flow)
T_i = integral time (reset time) (sec)
T_r = rise time (time to first reach setpoint) (sec)
T_u = ultimate period (sec)
λ = closed loop time constant (sec)
θ_i = implied total loop dead time (sec)
θ_o = original total loop dead time (sec)
τ_o = open loop time constant in self-regulating process (sec) (negative feedback)
τ_f = signal filter or attenuating volume time constant (sec)
τ_L = load disturbance time constant (sec)
τ_p = primary process time constant (sec)
Δt_x = controller execution time (sec)

Appendix D
Discussion

Appendix D
Discussion

The following are questions and answers posted on the Emerson Exchange 365 Mentoring Discussion Website as part of the ISA Mentor program.

D.1 Danaca Jordan's Question on Tuning Usually Inactive Loops

I am trying to improve the control of a differential pressure controller (PDIC) that is only turned on when things go wrong on a particular product, so it is only active for a couple of unplanned hours at a time. Any suggestions for using limited historian or Insight data to improve tuning?

Greg's Answer

You can estimate the dead time from the time it takes the PID PV to change direction after a change in PID output direction when the PDIC is active. For a large change in PID output, you can approximate an integrating process gain as the new rate of change minus the old rate of change (including sign) in %/sec divided by the % change in PID output yielding a gain in %/sec/% or 1/sec. You can then use lambda tuning rules for integrating processes. I suggest a lambda equal to 3x the dead time to help account for nonlinearities. Even if you cannot estimate the integrating process gain, you can check the reset setting based on just knowing the dead time. Most pressure and differential pressure loops have too much integral action (too small of a reset time in seconds or too large of a reset setting in repeats/minute).

I suggest you try using the auto tuner relay oscillation method for a quick better estimate of dynamics. You can set the change in output to be just 5x the valve dead band, check the integrating process option, and go for just 2 cycles. If the valve dead band is 0.5% (typical for a reasonably good valve) the amplitude in PID output for the relay method can be as small as 2.5%. The relay method should be able to keep the PV near the setpoint for small load upsets while you are running the test but the estimates will be affected. You can manually check the integrating process tuning settings using a lambda equal to the 3x the dead time.

The assumption here is that you have a process where the primary time constant is more than 4x the dead time making it a near-integrating process. If this is not the case, the reset time will be too large but this will be in the conservative direction.

D.2 Héctor Torres's Question on Resolution Limits

In different articles and books, you will normally talk about resolution limits. I am not sure if I totally understand the concept and all it entails. What should we understand in the application of the term and how will I identify if I am experiencing problems stemming from resolution limits? What kind of control issues will I have? How can I solve a resolution limit problem?

Greg's Answer

A resolution limit is the smallest input change that will cause an output change. When the input change exceeds this limit, the output change is an integer multiple of the resolution limit. Thus, the output response is a series of steps with a size that is an integer multiple of the resolution limit. In contrast when a change in input exceeds the threshold sensitivity limit, the change in output equals the change in input. The result is a step size that matches the input change. Resolution limits come from digital devices. However, the ISA standard for valve response testing expresses stick-slip as a resolution limit when really a threshold sensitivity limit is more at play where slip is about equal to the stick.

The 1980s vintage DCS had only 12 bits in the input cards. Since one bit is for sign, there were only 11 bits for resolution. If you divide 100% by 2 to the 11th power, you get a resolution limit of 0.05% which doesn't seem bad until you consider the practice at the time was to use wide range thermocouple input cards rather than transmitters to save on field hardware costs. These cards caused a series of steps of about 0.4°F that caused such severe bumps in the PID output from rate action that derivative could not be used on reactors where derivative is most needed to compensate for secondary time constants from thermal lags. Today's DCS has 16 bit input and output cards.

The standard signal input card for variable speed drives (VSD) last time I checked (2008) has only 8 bits which if there is one sign bit translates to a resolution of 0.78%. This is really lousy when you consider the resolution limit of a sliding stem valve, digital valve controller, and diaphragm actuator is close to 0.1%. Users need to specify a VSD input card with at least 12 bits. Why this is a special card shows the lack of understanding of resolution limits.

When a step change in the PV occurs, the derivative mode thinks the entire change occurred in one PID execution. Exothermic liquid reactors have a slow temperature response and use a PID gain greater than 10 and a rate time greater than 120 seconds to prevent a runaway. Since the step from a resolution limit is divided by the PID execution time and multiplied by the product of the PID gain and rate time, a 0.4°F step can cause a full scale bump in the PID output.

Resolution and threshold sensitivity limits both cause limit cycles if there is any integral action in the process or the controller. The amplitude depends upon the process gain. For pH control on the steep part of a titration curve the process gain is so large, the limit cycle can violate pH limits. Such pH processes are very sensitive to valve resolution and threshold sensitivity limits.

Most sensors and analyzers have a threshold sensitivity limit larger than the resolution limit in today's smart transmitters. The wireless default trigger level is really a threshold sensitivity limit. The default update rate helps prevent a limit cycle if the trigger level is set too large.

The best solution is to use transmitters and valves with the best resolution and threshold sensitivity. Limit cycles and bumps in the PID output from rate action can be eliminated by the use the enhanced PID with external reset feedback.

D.3 Héctor Torres's Question on Setpoint Lead-Lag

One of the options you recommend for eliminating overshoot at setpoint changes is using a Setpoint Lead-Lag setup. I cannot visualize the concept in my mind. In a 2 Degrees of Freedom PID Structure (2DOF) you will have an idea of what beta and gamma do and where they are in the execution path.

For the Setpoint Lead-Lag setup, I do not comprehend where the lead and where the lag act; in other words, what is led and what is lagged?

Greg's Answer

To get a Setpoint Lead-Lag, the PID setpoint would be passed through a Lead-Lag function block. A Lead-Lag block is normally used for dynamic compensation of feedforward signals. For setpoint changes we are using this block to moderate the step in the PID output from the proportional mode and the bump from the derivative mode and to prevent immediately driving the PID output past its final resting value. The tuning settings that reduce the peak and integrated errors from unmeasured step disturbances to the process input maximize the step and bump in the PID output for setpoint changes making the approach to setpoint faster but causing excessive setpoint overshoot. Preventing overshoot is important in many applications and reducing abrupt movements in the PID output that could upset other loops and systems is important in gas volumes and utility headers.

In the frequency domain, the lead is a zero and the lag is a pole. The Laplace transform shows that the numerator term for the zero has the same form as the denominator term for the pole. If the lead time equals the lag time, the numerator (zero) and denominator (pole) cancel out and the output equals the input. If the lead time is zero, the output is simply the filtered input where the lag time is the filter time. If the lead time is nonzero but less than the lag time, there is a small step in the block output followed by the exponential approach to the final value of the input set by the lag (filter) time. If the lead time is greater than the lag time, there is a spike in the output that is greater than the input followed by an exponential decay to the final value of the input set by the lag (filter) time.

If the lead time is zero and the lag time is set equal to the PID reset time, the response of the PID should be similar to a PD on PV and I on Error structure or a 2 Degrees of Freedom (2DOF) structure with beta and gamma both equal to zero. This will prevent overshoot and provide a smooth and gradual change for both the PID output and PV. The time to reach setpoint (rise time) is much longer.

For bioreactors, gas furnaces and reactors, and plug flow systems (e.g., pipes, static mixers, extruders, sheet lines) a smooth approach with no overshoot is much more important than rise time. Bioreactor cells are very sensitive to abrupt changes and to a temperature or pH higher than optimum. Gas and plug flow volumes have no primary process time constant to filter out abrupt changes in PID output and reach a final value relatively quickly without having to make an abrupt change in the PID output.

For large, well-mixed liquid volumes, the increase in rise time for setpoint changes in startup, transitions, and batch operations can appreciably reduce process capacity. The response of these volumes for continuous operations is near-integrating and for batch operations is essentially true integrating. For both types of operations on these volumes, lambda integrating process tuning rules are used. The inherent approach to setpoint is very slow due to a large primary process time constant or slow integrating process gain making it necessary to have a large sudden change in the PID output to get to significantly different setpoint in minutes rather than hours.

If the lead time is set equal to 25% of the lag time, and the lag time is set equal to the reset time, you get a response with a good compromise between setpoint overshoot and rise time. Similar results can be obtained by a 2DOF structure with the beta equal to 0.5 and the gamma = 0.25. However, in my test results, the 2DOF structure with any gamma setting > than zero resulted in some short term higher frequency oscillations in the PID output whereas with the Lead-Lag there was simply one bump. The 2DOF oscillations disappear quickly and may not exist for less aggressive tuning than what I used.

Since adding a Lead-Lag function block to all setpoint changes requires more effort than simply changing the PID structure to 2DOF, the 2DOF option may be better. Since you can achieve many of the other PID structures that have integral action by a simple change of the beta and gamma, one could say for flexibility, the 2DOF structure may be the future way to go.

You can go to DeltaV Books Online to find out more about the Lead-Lag block and PID structures.

D.4 Héctor Torres's Question on Measurement Lag

How important is it to determine the lag caused by my measurement system?

At a bump test I will note the associated delay time and the change in my process variable; any lag in my measurement will already be acting on the observed dynamics. Since the computed tuning parameters will be based on these, they will be already accounting for such measurement lag; will it make a difference determining how much is coming from such source?

When will it be important to determine the measurement lag?

Greg's Answer

Excellent question that really made me think. The motivation to identify the size of the measurement lag is in the knowledge whether the lag is causing a significant deception in the view or delay in the correction of process variability. Excessive lag in many cases can be fixed by an adjustment, replacement of the sensor, or redesign of the installation. The solution is often within the responsibility and capability of the automation engineer. Since the measurement is the window for analyzing and controlling the process, your knowledge and solution can be essential.

Deception occurs when the measurement lag is greater than the ultimate period of the loop. The large measurement lag acts as a filter attenuating process variability. The measurement amplitude of an oscillation is less than the actual amplitude. Doing the wrong thing (i.e., making the measurement lag larger), makes the trend chart look better in terms of less variability. If the measurement lag becomes the largest time constant in the loop, an increase in the lag enables an increase in the PID gain furthering the deception that the measurement lag is beneficial. Normally, the reset time cannot be decreased and may in fact need to be increased to stop a reset cycle. So the clue may be a slower more oscillatory response. The ratio of the measured to actual process variable amplitude is the period of the oscillation divided by 6.28 times the lag. This simple relationship enables one to estimate the degree of deception and the actual process variable amplitude from the observed amplitude for a given oscillation.

A noticeable effect of measurement delay occurs when the measurement lag is more than 10% of the PID reset time. The effect is seen in terms of an increase in the measured peak and integrated error for a load disturbance and the rise time for a setpoint change. The effect becomes intolerable if the additional delay from the lag increases the total loop dead time to the point where the PID must be re-tuned. This occurs when the additional dead time from measurement lag increases the total loop dead time to be greater than the implied dead time based on the tuning. The result is often an oscillatory response from excessive integral action. The implied dead time is ½ and ¼ the sum of lambda plus the original loop dead time (dead time with no measurement lag) for self-regulating and near or true integrating processes, respectively. If the measurement lag is smaller than the process time constant, the additional dead time comes from the larger measurement lag itself. If the measurement lag is greater than the process time constant, the additional dead time comes indirectly from a greater fraction of other time constants becoming effectively dead time. A clue as to whether the measurement lag has increased is a process variable that oscillates with a larger period. Due to attenuation/deception, the amplitude may look less.

If the measurement lag becomes the largest time constant in a loop for an integrating process or even worse for a runaway process (e.g., highly exothermic reaction), the deterioration is huge. Setting the PID rate time equal to a large measurement lag is important. For runaway processes a large measurement lag can make the maximum allowable PID gain approach the minimum allowable PID gain needed to prevent the process from accelerating and reaching a point of no return. The window of allowable PID gains can close in a runaway process due to a large measurement lag (a serious safety risk).

How do you identify the measurement lag? How do you know the primary or secondary time constant identified by software is in the measurement, process, or final control element (e.g., control valve or variable frequency drive)? There are few easy cases. The PID process variable filter time and transmitter damping should be checked and reduced if they are causing deception or deterioration per above guidelines. The settings should be minimized if the settings are larger than 10% of the reset time. There is a chicken and egg possibility, in that the reset time may be large due to a large measurement lag.

If a step change in flow can be made to a process input, the open loop response of the measure variable will exclude the effect of the final control element lag from its rate limited exponential response and inherent time constant (another story). Also, the time constant for a small sliding stem valve or a variable frequency drive with insignificant rate limiting in the drive setup is negligible for small changes (e.g., <2%) in PID output. This still leaves us with the quandary of how to know if the primary or secondary time constant identified is in the process or the measurement.

For fast process where there is negligible back mixing, the process time constant is small. The process time constant is less than 1 second for gas flow reactor and furnace composition and temperature control by manipulation of feeds or fuel, liquid or polymer pressure and flow control, and inline pH control. A primary or secondary time constant much larger than 1 second is most likely a measurement lag. The electrode time constant is about 2 to 6 seconds for a new clean electrode and reasonable velocity (e.g., > 5 fps). An aged or coated glass electrode can increase the measurement lag from a few seconds to minutes, even hours. A loosely fitted temperature sensor in a thermowell or even worse a thermocouple in a ceramic protection tube will cause the measurement lag to increase from seconds to minutes due to the air gap or ceramic tube acting as an insulator.

Transmitters and electrodes can be tested in the shop to determine the measurement lag excluding any installation effects. The transmitters and electrodes should not be cleaned, which would eliminate the effect of coatings. Electrodes are inserted into buffer solutions or even better process samples at operating conditions. Temperature sensors still in their thermowells can be inserted into temperature baths. Velocity affects the electrode and thermowell response, so a stirrer should give about the same velocity as in the process installation. The measurement lag is ½ the 86% response time assuming the measurement delay is negligible. Waiting for a more complete response (e.g., 95% response time) especially for pH electrodes leads to less relevant and more inconsistent results.

Appendix E Enhanced PID Analyzer Applications

Appendix E
Enhanced PID Analyzer Applications

Introduction and Overview

It is widely recognized that wireless measurements considerably reduce installation and maintenance costs. However, it is less recognized that the portability of wireless offers the ability to optimize the measurement location for the fastest, most representative process response and to demonstrate/prototype process control improvement for finding and justifying permanent solutions.

The wireless update time termed "default update rate" (refresh time) is the time interval for periodic reporting. The wireless update sensitivity termed "trigger level" is the minimum change in measurement value for exception reporting. To save power, the transmitter goes to sleep and wakes up periodically to check if the change in sensor value from the last value transmitted is larger than the trigger level. When the change exceeds the trigger level or the time since the last communication exceeds the default update rate, then the value is transmitted. The time interval between periodic checks of the sensor is the "triggered update rate" (wakeup time). If only periodic reporting is available, this time interval is the default update rate. Increases in the default update rate and trigger level settings reduce the number of transmissions, which increases battery life. The enhanced PID eliminates the ramps, limit cycles, and spikes from large values of these settings facilitating an increase in battery life.

For slow processes where a significant process response takes minutes to hours to develop, such as composition, level, and temperature in large volumes, wireless measurements are fast enough for closed loop control using traditional PID. The primary question for these applications is what is the potential detrimental effect of a loss of update and poor sensitivity? The enhanced PID protects against communication, sensor, valve, and analyzer failures, and eliminates the cycling from wireless exception reporting, sensor, valve, or analyzer sensitivity limits for these and other loops. Thus, the enhanced PID improves the reliability and reduces the variability for all loops with and without wireless measurements. Additionally the enhanced PID extends the applicability of wireless measurements and analyzers to processes with a response time faster than the update time.

While the enhanced PID offers advantages for all loops, wireless measurements should not be used where the process can cause product degradation, shutdowns, equipment damage, or an unsafe condition faster than the triggered update rate (exception reporting) or default update rate (periodic reporting) of the wireless transmitter. Examples of loops that can get into trouble too fast for even the fastest wireless update time are sheet, compressor, and turbine speed control, and incinerator and pipeline pressure control. Slow wireless default update rates, such as 30 and 60 seconds, can be a problem for vessel or header pressure control since the pressure may approach alarm, trip, or relief settings within the default update rate.

In a traditional PID, the integral and derivative modes are computed each execution of the PID block.

The PID algorithm uses the execution time in the integral and derivative mode calculations. Thus, the reset and derivative contribution calculated by traditional PID may not be appropriate when used with a wireless measurement where the default update rate is significant compared to the process response time. In such cases, the traditional PID will ramp the controller output through continual integral action even though the actual measurement has not changed. The PID is acting on old information. The detrimental effect of integral mode acting on old information is greatest if the process response has largely responded before the next update, which occurs when the process response is faster than the update time or the measurement value has not been updated because of device or communication failure. The update time of wireless measurements on fast processes, such as flow and pipeline temperature and pH loops, and the update time for analyzers in many applications are slower than the process response. An update failure occurs for loss of transmission, sensor failure, a stuck valve, and a sample or multiplexer system failure. Furthermore, when the wireless measurement is communicated using exception reporting, the ramp of the controller output between updates due to a sensitivity limit results in a perpetual equal amplitude oscillation called a limit cycle even though there are no load disturbances.

When there is an update, the traditional PID considers the entire change in the measurement value occurred within the PID execution time for the derivative mode calculation. The result when the PID is executed faster than the wireless default update rate is a spike in controller output whose size increases with the movement of the process between updates, which is greatest for a fast process or for recovery from an update failure.

The ramp, limit cycle, and spike from a traditional PID inflict a disturbance upon the loop and possibly other loops. Thus, a traditional PID in automatic mode can cause process variability even if there are no disturbances.

The enhanced PID computes the integral and derivative mode contributions to the controller output when there is a measurement update and uses the elapsed time between updates in its calculations. Thus, the enhanced PID only acts on new information and considers the observed change in the measurement to have occurred not in just the last PID execution time but over the elapsed time. The key functional features of the enhanced PID are:

- PID responds immediately to setpoint, feedforward, and manual/remote output changes.
- PID responds immediately to process variable changes.
- PID action is suspended if there are no changes (integral action stops).
- PID action uses elapsed time instead of PID execution time in derivative calculation.
- PID action computes positive feedback filter exponential response using elapsed time.

If the dead time from a discontinuous signal whether a digital device or at-line or online analyzer is much greater than the open loop time constant, the enhanced PID will see the full process response in the next update. For these cases, the controller gain can be set equal to the inverse of the controller gain to provide considerably tighter control. Variation in the update time does not affect the tuning. Consequently, the enhanced PID is particularly valuable for closed loop control using off-line analyzers where the time between lab results is large and extremely variable.

When there is a loss of communication with the measurement or final control element (e.g., valve, damper, or variable speed drive), the enhanced PID provides no further reset action. The enhanced PID waits for new information whereas the traditional PID output ramps to an output limit. When communication is restored, the enhanced PID acts on the new information only. The traditional PID sees the effect of the ramp to the output limit that occurred during the loss of communication or loss of valve travel. Additionally, the traditional PID considers the entire observed change to have occurred within the last execution of the PID instead of the time duration of the failure. The result is a bump from gain action and a spike from rate action in the traditional PID output.

The enhanced PID executes as fast as a traditional PID independent of the wireless default update rate or analyzer cycle and sample time. As a result, enhanced PID is ready to immediately act on changes in setpoint, mode, remote output, feedforward, and tuning. For a proportional gain that is the inverse of the process gain, the setpoint response is particularly impressive in that the controller output goes to the value needed to reach setpoint within the PID execution time. For a fast loop, the process is already at the setpoint before the next update.

Secondary loops are designed to be faster than primary loops for cascade control. If a wireless measurement is used in a secondary loop, the enhanced PID can prevent the secondary loop response from becoming too slow. The fast load response of the enhanced PID helps the secondary loop reject secondary disturbances. The fast setpoint response of the enhanced PID helps the primary loop reach its setpoint faster and reject primary loop disturbances.

Analyzers with sample systems offer a higher level of control. Ultimately you want to know and control is the composition in a process stream

However, the use of analyzers for closed loop control is rather limited due to the long update time, excessive interruptions and extraneous values, and low sensitivity. Analyzer sensitivity is generally not as good as flow, pH, pressure, and temperature measurement sensitivity. The analyzer sensitivity limit is the result of interferences, analysis method, and sensor sensitivity. Analyzer update time, poor sensitivity, and noise results in a step response that precludes the use rate action even if the process time constant is large. The enhanced PID eliminates most of these concerns and enables more analyzers to go on closed loop control.

If there is a resolution instead of a sensitivity limit, where the response is quantized so that the increment or decrement in the measurement is an integer multiple of the resolution limit, the steps in the response never put the measurement at the setpoint. Analyzers with sample systems frequently have resolution limits due to analysis methodology and minimum dilution and reagent dose sizes. The enhanced PID will eliminate the cycling from reset action and the spikes from rate action for both resolution and sensitivity limits in the measurement. The enhanced PID can also eliminate the limit cycles from valve stick-slip and backlash (valve sensitivity limits).

The enhanced PID is available starting in DeltaV v11. The enhanced PID option can be enabled under the FRSIPID_OPTS for the PID block in control studio (Use PIDPlus). This PIDPlus option turns on the Dynamic Reset Limit option.

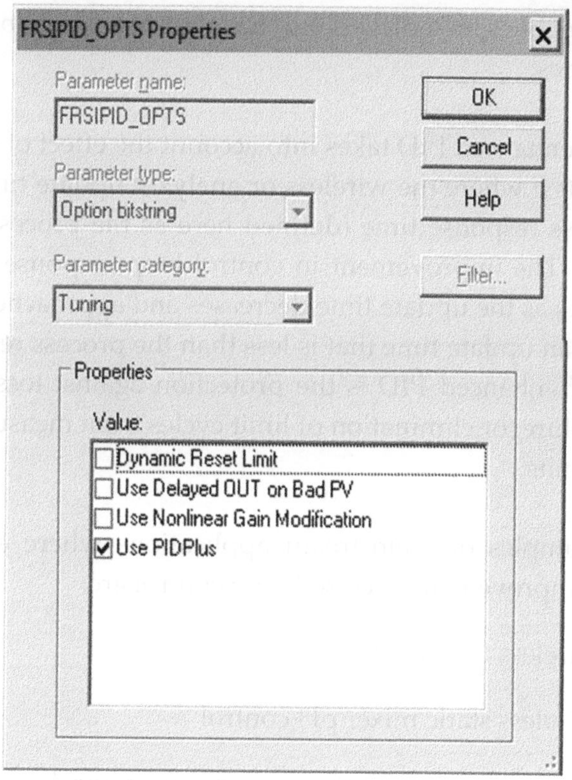

Figure E.1 – Enhanced PID Enabled in FRSIPID_OPTS for PID Block

Target Applications

While the enhanced PID can be used in any application to reduce process variability and provide protection against measurement and valve failures, the greatest benefits are seen for loops where:

- Update time is larger than the process response time.
- Update time is significant and variable.
- Update sensitivity limit is larger than the desired process precision.
- Update failures can occur (e.g., communication and electrode failures).
- Measurement resolution or sensitivity is poor causing limit cycles.
- Valve exhibits significant stick-slip and backlash causing limit cycles.
- Variable speed drive setup has significant deadband causing limit cycles.

Since the enhanced PID takes into account the effect of update time, the benefit is greatest where the wireless or analyzer update time is much larger than the process response time (defined here as the process dead time plus time constant). The improvement in control loop response can be significant but diminishes as the update time decreases and approaches the process response time. For an update time that is less than the process response time, the larger benefit of enhanced PID is the protection against loss of communication or device failure, or elimination of limit cycles from measurement and valve sensitivity limits.

Some examples of mainstream applications where enhanced PID provide greatest improvement in closed loop control are:

- Wireless flow control
- Wireless static mixer pH control
- Wireless desuperheater temperature control
- Pulp and paper stock consistency control
- Cross directional (CD) control of sheet thickness

- At-line analyzer control of pipeline composition
- At-line analyzer control of plug flow reactor product composition
- Off-line analyzer control of almost any unit operation
- Examples of mainstream applications with less but still significant improvement in closed loop control are:
- Wireless vessel pressure control
- Wireless extruder temperature control
- Wireless jacket temperature control
- Wireless heat exchanger temperature control

Test Setup

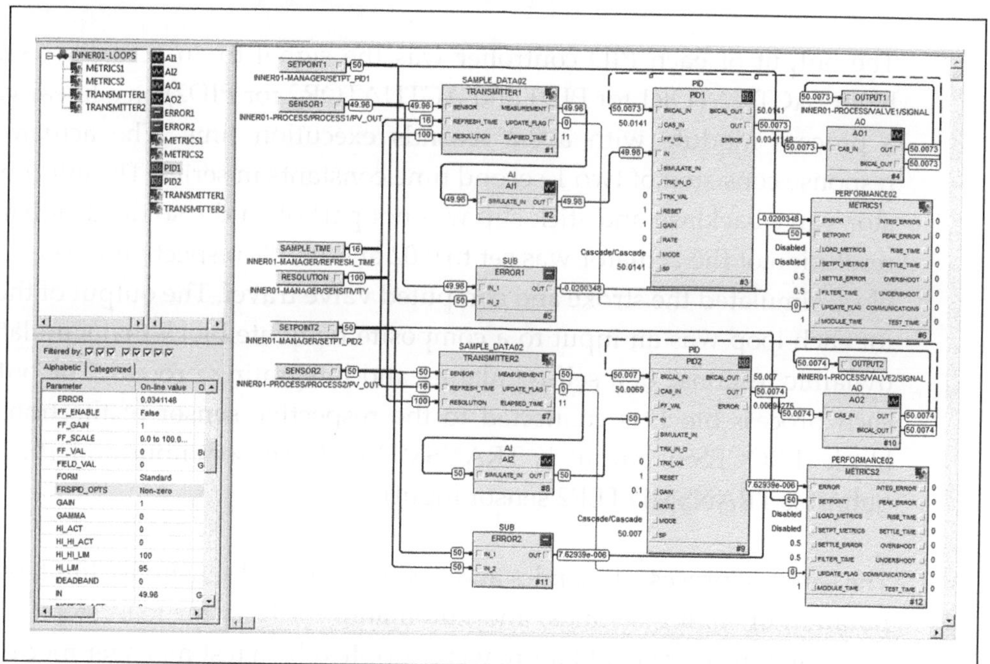

Figure E.2 – Virtual Plant Test Setup for Comparison of Enhanced and Traditional PID

A DeltaV v11 module with 1 second execution time was set up with two PID blocks. The first PID block (PID1) had the enhanced PID option enabled in the FRSIPID_OPTS parameter whereas the second PID block (PID2) did not. The

PID analog inputs were provided by SAMPLE_DATA02 composite blocks that simulated the update time and sensitivity limits for wireless transmitters and analyzers (TRANSMITTER1 for PID1 and TRANSMITTER2 for PID2). A composite block (PERFORMANCE02) for performance metrics tallies the number of communications and computes the integrated absolute error (IAE) and peak error for load disturbances and the rise time, settling time, overshoot, and undershoot for setpoint changes

Many of the wireless devices on the market today support default update rates of 8, 16, 30 and 60 seconds. However, to maximize battery life it is assumed that the longest reasonable default update rate will be selected. The default update rates selected for the tests were 16 and 30 seconds for the fastest loops and 30 and 60 seconds for the slowest loops.

Lambda tuning was used for both PID controllers. Rate action was not used. In all of the cases studied except vessel temperature control, the PID2 gain had to be decreased to prevent excessive oscillations.

The output of each PID controller was the signal to an actuator composite block (ACTUATOR1 for PID1 and ACTUATOR2 for PID2) on a linear valve in a process module with a 0.2 seconds execution time. The actuator time response consisted of two 1-second time constants in series. The effect of actuator-valve backlash and stick-slip was not part of study, so the deadband and resolution of the actuator was set to 0.02 and 0.01% respectively. The actuator block simulated the stroke and computed valve travel. The output of the valve for each loop was an input to a composite template block that simulated the dynamic response of a self-regulating and integrating process. The output of each process block is connected to the respective sensor of the transmitter block (PROCESS1 output is TRANSMITTER1 sensor input and PROCESS2 output is TRANSMITTER2 sensor input).

The setup of process dynamics and update time and sensitivity limit, the computation of tuning settings, and automation of tests for load steps, setpoint steps, and measurement failure were supplied by a test manager module with an execution time of 2 seconds. The manager module also turned on the performance metrics and the valve travel accumulation during the tests for load upsets and setpoint changes.

Discussion of Test Results

The variables in the trend charts are as follows:

1. Enhanced PID setpoint
2. Enhanced PID measured process variable (update set by wireless device or analyzer)
3. Enhanced PID actual process variable (instantaneous update)
4. Enhanced PID output
5. Traditional PID setpoint
6. Traditional PID measured process variable (update set by wireless device or analyzer)
7. Traditional PID actual process variable (instantaneous update)
8. Traditional PID output

The size of the step change in load or setpoint for each loop was 10% except for vessel pressure where the step size was reduced to 5% due to the small vessel size. The actual process response included two 1-second actuator lags besides the stated process delay (dead time), process lag (time constant), and process gain.

Flow Loop

Flow loops are the most common loop in the process industry. The process dynamics are fast with a response time less than 2 seconds. The use of a wireless measurement with a traditional PID will oscillate severely unless the loop is detuned. The approach taken in the tests was to severely reduce the controller gain approaching an integral-only type of control. The reset time was kept at its original value to help eliminate offsets for ratio control.

The enhanced PID tuning could be made more aggressive than the tuning for a flow loop with a wired measurement. The controller gain could be increased to be the inverse of the process gain. For step setpoint and load changes, this enabled the enhanced PID to make a single almost exact correction. For a setpoint change, the correction occurs within one module execution time. How soon the response fully appears in the actual flow measurement is limited only by the valve response time (two 1-second lags in series), the process dead

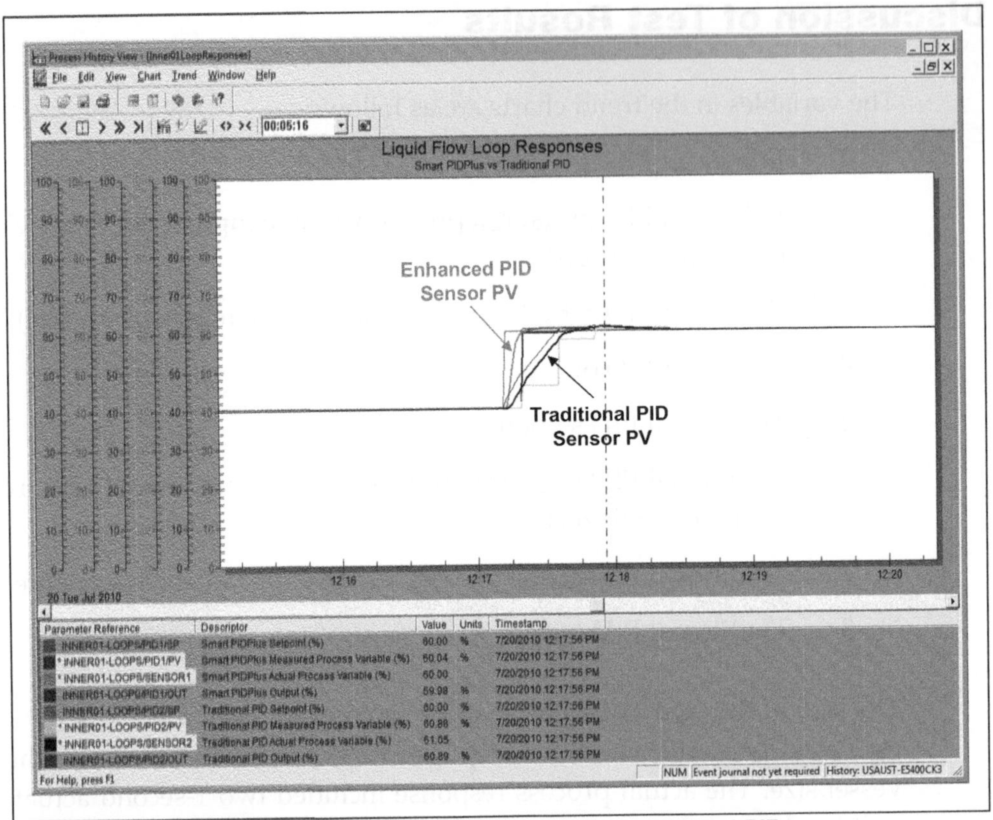

Figure E.3 – Flow Setpoint Response of Enhanced and Traditional PID (Update Time = 16 sec)

time (0.5 seconds), and the process time constant (1 sec). However, this incredibly fast response is not apparent to the user because the measurement update is much longer. In the trend charts, the actual process measured process variable is plotted. The enhanced PID load response is much faster than a traditional PID even though the correction is delayed on the average by about half of the update time after the upset. The exact delay in the recognition of the disturbance depends upon when the disturbance arrives in the update time interval.

For a 16-second pH update time and load upsets, the enhanced PID reduced the integrated absolute error (IAE) by 45%. Since the flow loop has a loop dead time that is not significantly less than the process time constant, the peak error of the enhanced PID and traditional loop for a load disturbance is the full open loop error (error with loop in manual). For a 16-second pH update time and setpoint changes, the enhanced PID reduced the rise time (time to

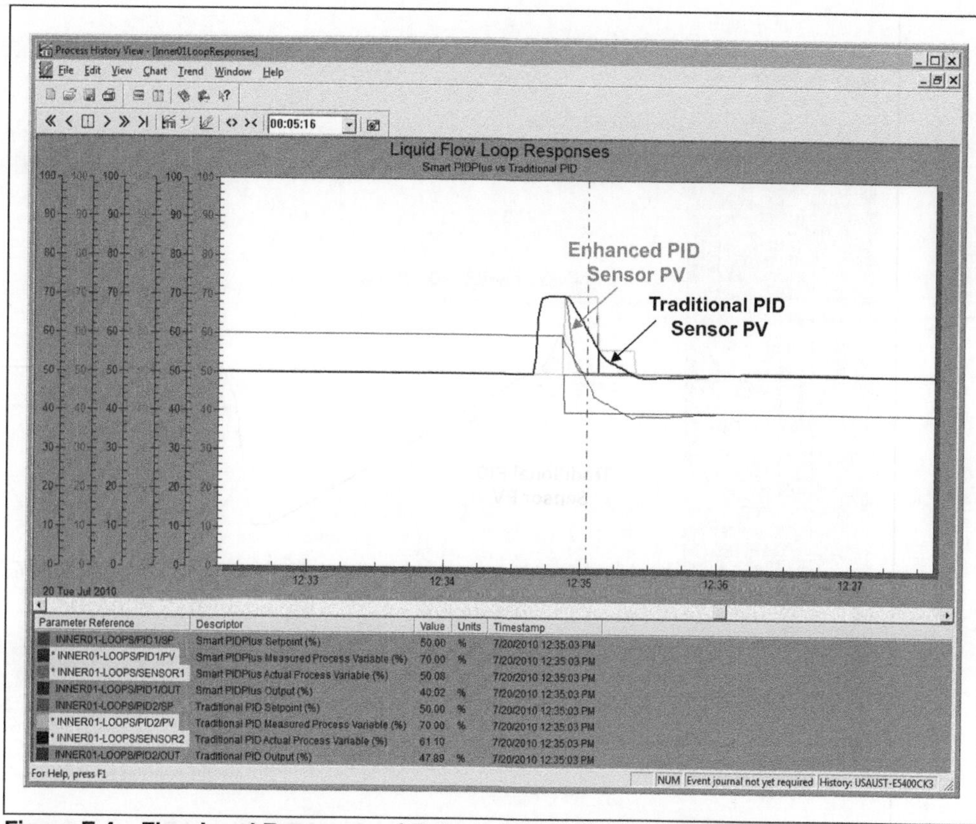

Figure E.4 – Flow Load Response of Enhanced and Traditional PID (Update Time = 16 sec)

reach setpoint) by 69%, the settling time (time to settle at setpoint) by 56%, and the overshoot by 55%.

If there is an update failure, the traditional PID will ramp towards an output limit. When the measurement is restored, the traditional PID sees the huge self-inflicted upset and overshoots the setpoint in its recovery. The enhanced PID corrects for upsets just before the failure and waits for the next update. The bump to the process is minimal if there are no upsets during the failure.

Static Mixer pH Loop

The enhanced PID dramatically improves the response of cascade loops by enabling an immediate response of the secondary loop to changes in its setpoint. Since the static mixer process dead time (6 sec due to injection delay) and process time constant (6 sec due to electrode lag) is much less than the default update rate (60 sec), the enhanced PID controller gain can be

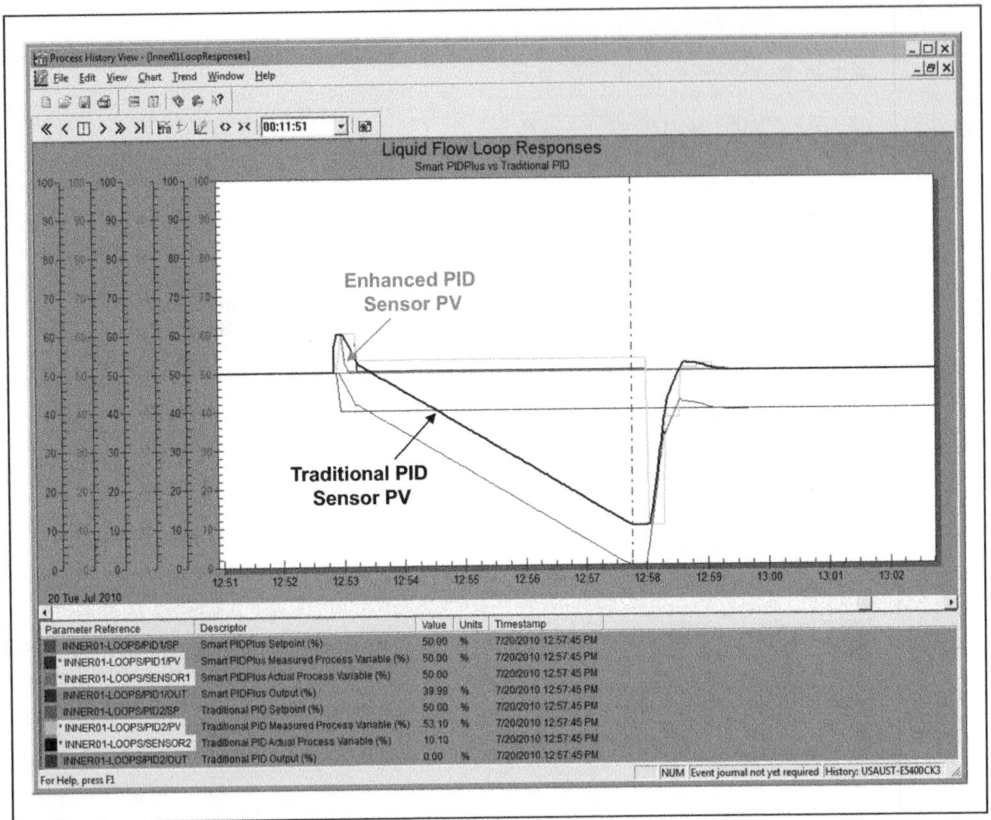

Figure E.5 – Flow Failure Response of Enhanced and Traditional PID (Update Time = 16 sec)

increased. With both the secondary flow and primary pH loop controller gain equal to the inverse of the process gain, the response to pH setpoint changes and load disturbances is impressive. In contrast, the traditional PID controller gains for both loops were a factor of ten lower to prevent oscillations. If the default update rate of the secondary loop is not about 4 times faster than the primary loop, the enhanced PID controller gain must be detuned to stabilize the loops since the oscillations from the secondary loop can be amplified by the primary loop. For a primary pH loop default update rate of 60 seconds, the secondary flow default update rate needs to be 16 seconds or faster.

For a 60-second pH update time and load upsets, the enhanced PID reduced the IAE by 40%. Since both the flow and pH loop have a process dead time that is not significantly less than the process time constant, the peak error of the enhanced PID and traditional loop for a load disturbance is the full open loop error (error with loop in manual). For a 60-second pH update time and setpoint changes, the enhanced PID reduced the rise time by 60%, the settling time by 84%, and eliminated the overshoot that was 40% of the setpoint change.

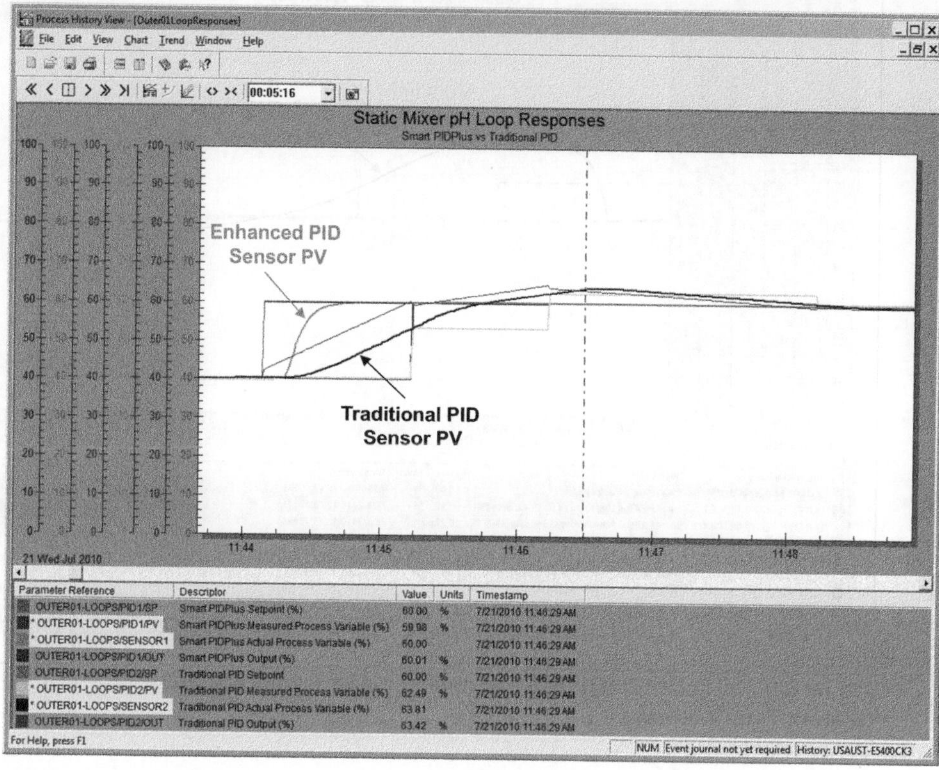

Figure E.6 – pH Setpoint Response of Enhanced and Traditional PID (Update Time = 60 sec)

Vessel Pressure Loop

The vessel pressure loop has an integrating response with a short process dead time (1 sec) and a relatively fast integrating process gain (0.1 %/sec/%) due to a small vessel volume or narrow measurement span. Even with a relatively fast default update rate of 16 seconds, control is challenging. The setpoint and load change for this test was 5%, whereas for all other tests it was 10%.

For a 16-second update time and load upsets, the enhanced PID reduced the IAE by 52% and the peak error by 30%. The reduction in peak error is important to prevent activation of relief devices and Safety Instrument Systems (SIS). For a 16-second update time and setpoint changes, the enhanced PID

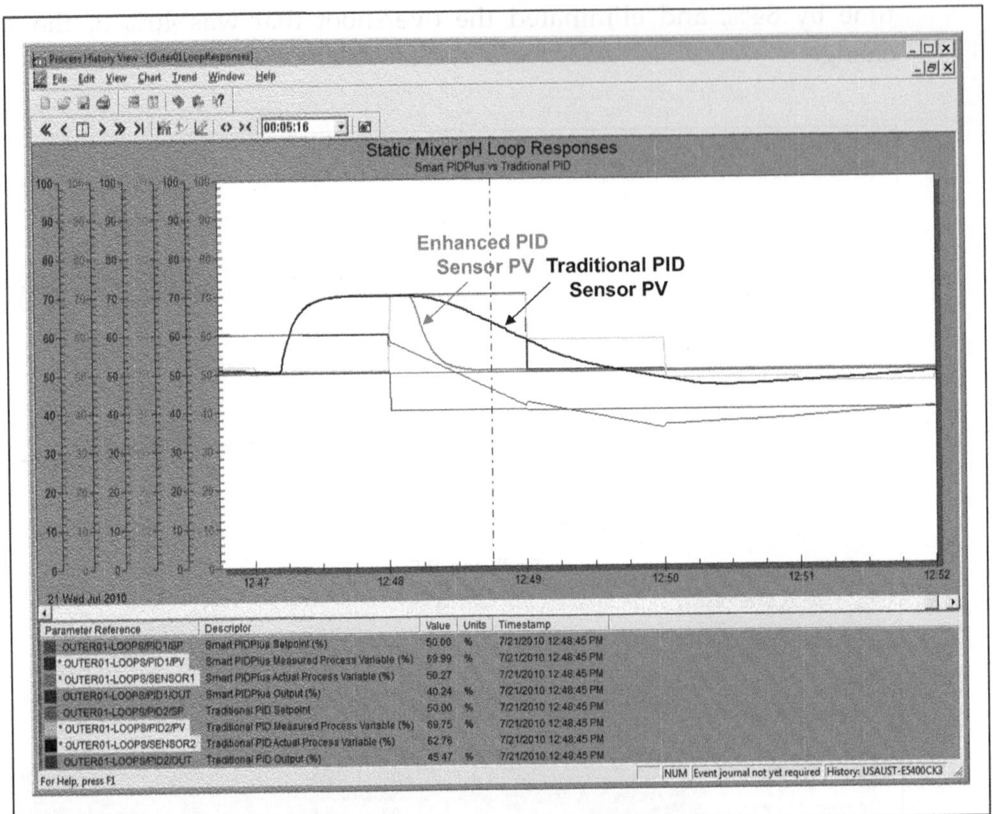

Figure E.7 – pH Load Response of Enhanced and Traditional PID (Update Time = 60 sec)

reduced the rise time by 39% and the settling time by 45%. However, the overshoot of the enhanced PID was 35% larger. Setpoint rate of change limits could be used to reduce the overshoot but this would increase the rise time.

Extruder Temperature Loop

For a 30-second update time and load upsets, the enhanced PID reduced the IAE by 18%. Since the extruder has a process dead time larger than the process time constant, the peak error of the enhanced PID and traditional loop for a load disturbance is the full open loop error (error with loop in manual). For a 30-second update time and setpoint changes, the enhanced PID reduced the rise time by 19%, the settling time by 42%, and the overshoot by 44%.

An extruder may have some radial mixing but there is negligible axial mixing (back mixing). As a result, the process dead time is larger than the process time constant and process fluctuations with time are not smoothed out and

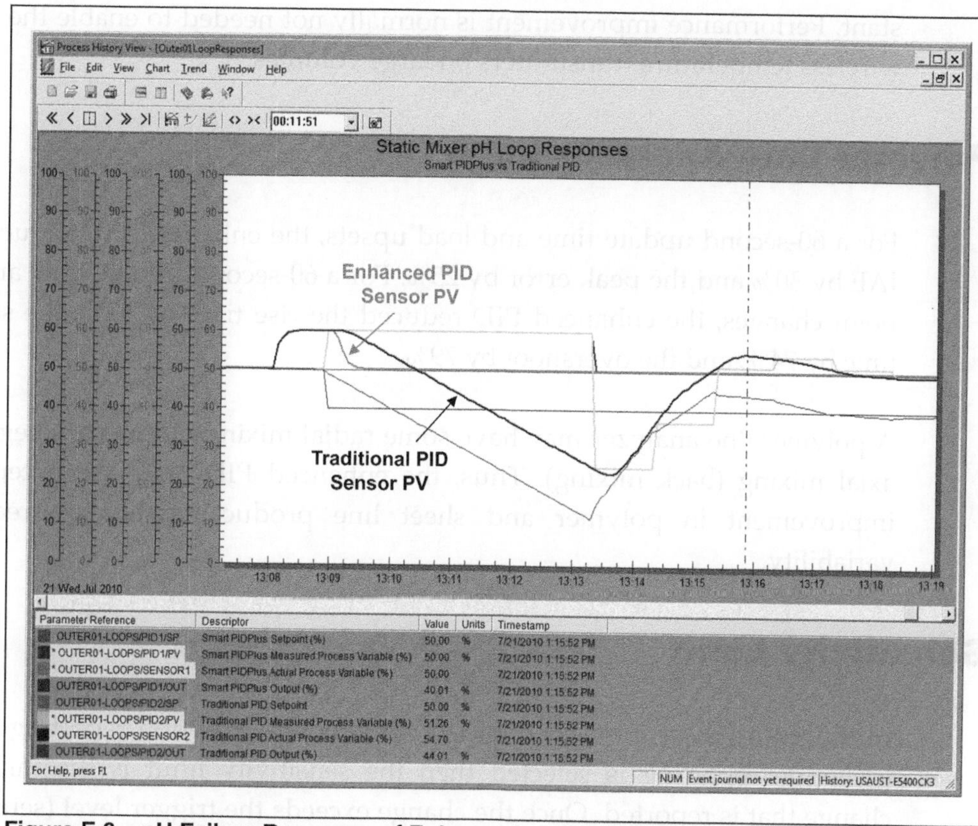

Figure E.8 – pH Failure Response of Enhanced and Traditional PID (Update Time = 60 sec)

appear in the product. Consequently, the ability of the enhanced PID to eliminate process variability directly translates to improved product quality.

Vessel Temperature Loop

For a 30-second update time and load upsets, the enhanced PID offered no improvement in the IAE or peak error since the process time constant is larger than the default update rate. For a 30-second update time and setpoint changes, the enhanced PID increased the rise time by 19%, but reduced the settling time by 27%, and decreased the overshoot by 31%. The enhanced PID could have been tuned for a faster rise time by increasing the controller gain. The vessel studied was exceptionally small. The performance of temperature loops is excellent on the larger vessel sizes typically seen in industrial plants if noise, sensitivity, and resolution limits are not significant and the controller is tuned with the high gain permitted by the extremely large process time constant. Performance improvement is normally not needed to enable the use of wireless temperature transmitters on large volumes.

Polymer Line Analyzer Loop

For a 60-second update time and load upsets, the enhanced PID reduced the IAE by 30% and the peak error by 25%. For a 60-second update time and setpoint changes, the enhanced PID reduced the rise time by 52%, the settling time by 74%, and the overshoot by 79%.

A polymer line analyzer may have some radial mixing but there is negligible axial mixing (back mixing). Thus, the enhanced PID offers an exceptional improvement in polymer and sheet line product quality by reducing variability.

Sensitivity Limit

In the future most wireless devices will support exception reporting. When exception reporting is selected then the sensitivity limit is the minimum change that is reported. Once the change exceeds the trigger level (sensitivity limit) or the time since the last communication exceeds the default update rate, the full change and exact change is reported. For a traditional PID, a sensitivity or resolution limit whether due to wireless exception reporting, sensor, valve, or analyzer sample and processing system, will cause a continuous oscillation termed a limit cycle. The amplitude of the square wave in the measurement is the sensitivity or resolution limit. Because the valve or measurement did not respond, traditional PID reset action continues to drive the output even though there is no process disturbance. When there is an update, the jump in the measurement goes beyond the setpoint and the ramp goes in the opposite direction. This ramping of the traditional PID output up and down continues indefinitely. Load disturbances and setpoint changes will temporarily disrupt the pattern. Since the enhanced PID integral mode only makes a calculation when there is an update, there is no ramping of the controller output. If there are no setpoint changes or load disturbances, the measurement draws a straight line. The actual process variable is closer to the setpoint than indicated by the measurement. The benefits are less process variability and less valve travel (longer valve packing life).

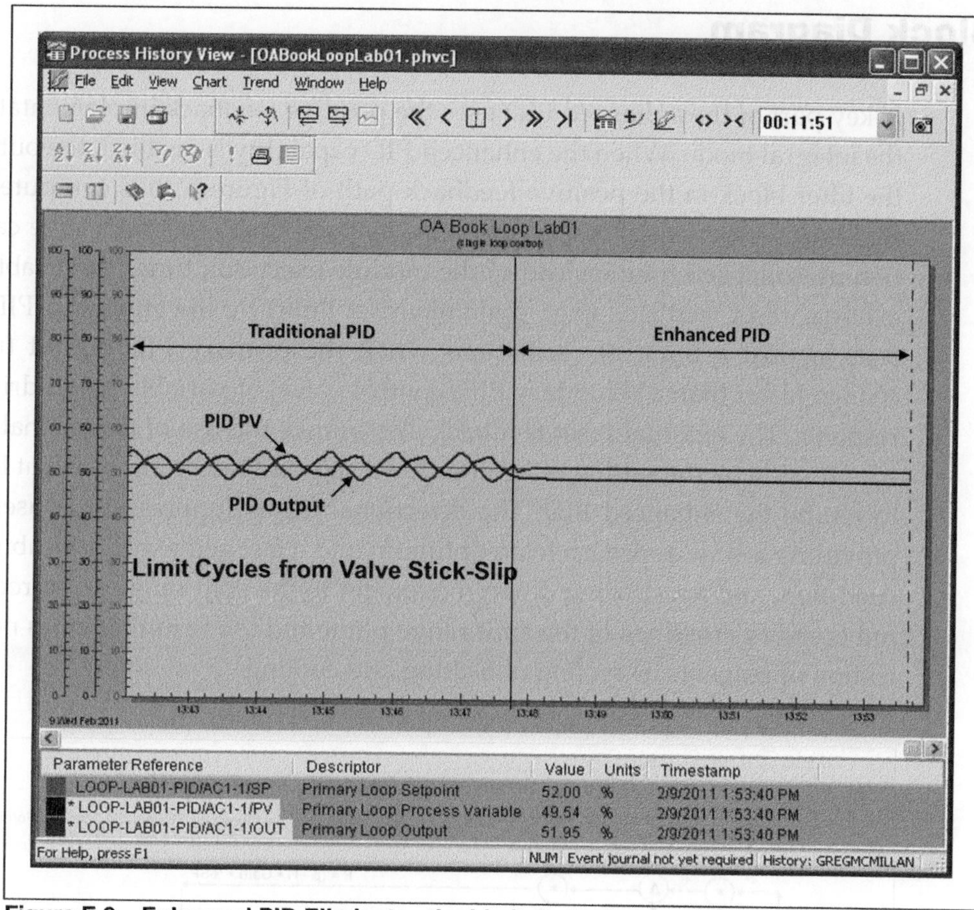

Figure E.9 – Enhanced PID Eliminates the Limit Cycle that Persists in a Traditional PID

The enhanced PID will also inherently eliminate limit cycles from control valve stiction and backlash and variable speed drive resolution limits. When the final control element is not responding, there is no process variable update unless there is a disturbance and consequently the ramping action of the integral mode is suspended. To ensure measurement noise does not trigger an update, any signal filtering needed should be judiciously done by transmitter damping. The wireless update trigger level should also be increased to screen out noise. To absolutely prevent an unnecessary update that would trigger PID action, a simple small sensitivity limit should be added to the process variable used by the PID block.

Block Diagram

A key part of the enhanced PID uses the positive feedback implementation of the integral mode. When the enhanced PID capability is enabled, the output of the filter block in the positive feedback path of Figure E.10 is computed as a first order exponential response for the elapsed time. The derivative calculation uses elapsed time instead of the module execution time. The enabling of external reset feedback (e.g., dynamic reset limit) by the enhanced PID protects against a burst of oscillations when the primary PID output tries to change faster than a secondary PID, control valve, or variable speed drive can respond. The external reset feedback also allows the use of directional move suppression by the setting of up and down setpoint rate limits without having to retune the enhanced PID. The directional move suppression is useful for providing a slow approach to an optimum and a fast getaway for an abnormal condition, and for slowing down movement to the split range point reducing unnecessary crossings of the split range point and the resulting cross neutralization of reagents or cycling of heating and cooling.

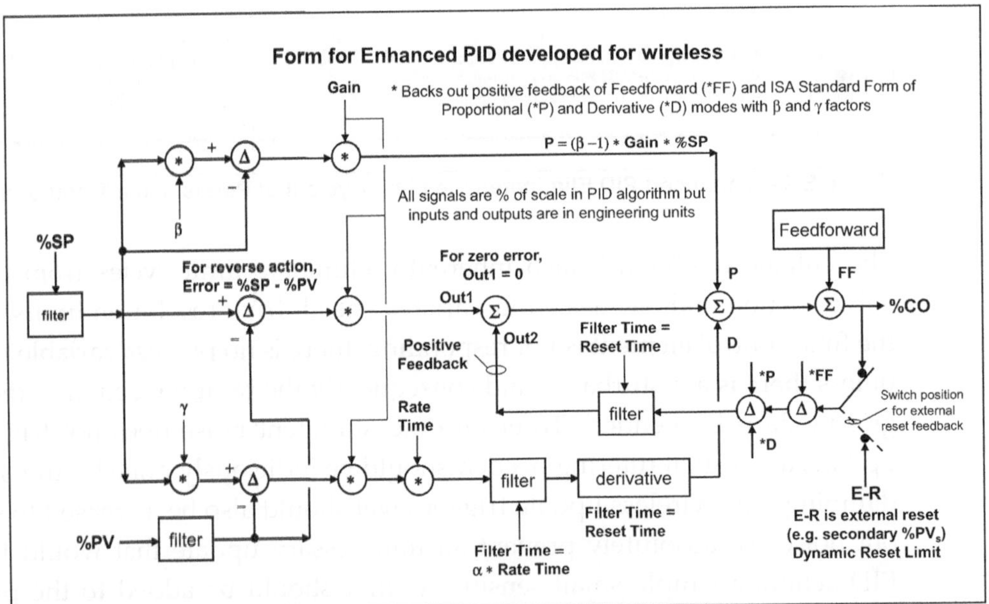

Figure E.10 – The Enhanced PID Uses the Positive Feedback Implementation of the Integral Mode as Shown Here for the ISA Standard Form

Summary of Test Results

Table E.1 – Summary of Setpoint Response Test Metrics

Process Type	Process Delay (sec)	Process Lag (sec)	Process Gain	Update Times (sec)	Overshoot PIDPlus , PID (%)	Rise Time PIDPlus , PID (sec)	Settling Time PIDPlus , PID (%)	Lambda PIDPlus , PID (sec)	Gain Setting PIDPlus , PID	Reset Setting PIDPlus , PID (sec)
Liquid Flow Inner Loop 1	0.5 sec	1 sec	1	16 sec 30 sec	0.5 , 1.1 0.0 , 2.3	9 , 29 10 , 23	25 , 57 10 , 98	2 , 2	1.0 , 0.1	1 , 1
Static Mixer pH Outer Loop 1	6 sec	6 sec	1	60 sec	0.0 , 4.0	38 , 94	38 , 244	8.5 , 8.5	1.0 , 0.1	6 , 6
Vessel Pressure Loop 2	1 sec	---	0.1 sec^{-1}	16 sec 30 sec	6.6 , 4.9 11.0 , 8.2	16 , 26 16 , 27	86 , 156 233 , 276	20 , 20	0.8 , 0.4	42.5 , 42.5
Extruder Temperature Loop 3	24 sec	6 sec	1	30 sec 60 sec	2.2 , 3.9 0.0 , 7.9	74 , 91 181 , 85	167 , 286 181 , 293	25.5 , 25.5	0.5 , 0.1	6 , 6
Vessel Temperature Loop 4	12 sec	60 sec	1	30 sec 60 sec	6.0 , 8.7 7.3 , 12.1	50 , 42 51 , 39	199 , 272 276 , 288	13.5 , 13.5	2.2 , 2.2	60 , 60
Polymer Line Analyzer Loop 5	1 sec	10 sec	1	30 sec 60 sec	2.5 , 0.0 0.4 , 1.9	25 , 128 39 , 81	63 , 128 39 , 147	2.5 , 2.5	1 , 0.2	10 , 10

The process delay in table does not include the two 1 sec time lags in the valve response

Table E.2 –Summary of Load Response Test Metrics

Process Type	Process Delay (sec)	Process Lag (sec)	Process Gain	Update Times (sec)	IAE PIDPlus , PID (%*min)	Peak Error PIDPlus , PID (%)	Valve Travel PIDPlus , PID (%)	Lambda PIDPlus , PID (sec)	Gain Setting PIDPlus , PID	Reset Setting PIDPlus , PID (sec)
Liquid Flow Inner Loop 1	0.5 sec	1 sec	1	16 sec 30 sec	6.5 , 11.9 9.2 , 19.2	20 , 20 20 , 20	30 , 33 30 , 65	2 , 2	1.0 , 0.1	1 , 1
Static Mixer pH Outer Loop 1	6 sec	6 sec	1	60 sec	45.3 , 75.5	20.0 , 19.7	--------	8.5 , 8.5	1.0 , 0.1	6 , 6
Vessel Pressure Loop 2	1 sec	---	0.1 sec^{-1}	16 sec 30 sec	15.3 , 32.0 34.8 , 53.2	15.1 , 21.5 31.5 , 32.0	35 , 40 82 , 79	20 , 20	0.8 , 0.4	42.5 , 42.5
Extruder Temperature Loop 3	24 sec	6 sec	1	30 sec 60 sec	34.9 , 42.5 69.5 , 67.9	20 , 20 19.6 , 18.8	33 , 46 28 , 60	25.5 , 25.5	0.5 , 0.1	6 , 6
Vessel Temperature Loop 4	12 sec	60 sec	1	30 sec 60 sec	17.2 , 16.6 27.4 , 40.6	8.9 , 8.0 13.9 , 14.3	50 , 97 59 , 212	13.5 , 13.5	2.2 , 2.2	60 , 60
Polymer Line Analyzer Loop 5	1 sec	10 sec	1	30 sec 60 sec	13.4 , 27.0 23.1 , 33.0	11.8 , 15.7 11.9 , 15.9	31 , 33 30 , 40	2.5 , 2.5	1 , 0.2	10 , 10

The process delay in table does not include the two 1 sec time lags in the valve response

Appendix F
First Principle Process Relationships

Appendix F
First Principle Process Relationships

General Implications

First principle relationships can define process cause and effects that can lead to improved controller tuning and performance by the selection of better tuning rules and process variables for scheduling of tuning settings. It also affects the choice of control valve trim and the feedforward design. The understanding of these relationships does not require a degree in chemical engineering but presumes just some understanding of common terms (e.g., heat transfer coefficient and area), relationships (e.g., ideal gas law), and physical concepts (e.g., conservation of mass and energy).

Equations have been developed from first principle relationships for the process gains, dead times, and time constants of volumes with various degrees of mixing. The results show that for well mixed volumes with negligible injection delays, the effect of flow cancels out for the controller gain if one of the following methods is used: lambda self-regulating rule where lambda is set equal to the dead time, or the reaction curve method. The effect of flow also cancels out for the reset time besides the controller gain if the process is treated as a "near integrator" and the lambda integrating tuning rule is used. This is because the flow rate cancels out in the computation of the ratio of process gain to time constant that is the "near integrator" gain. This ratio and "near integrator gain" are inversely proportional to the process holdup mass (e.g., liquid mass). However, for temperature control the effect of changes in liquid mass cancels out because a change in level increases the heat transfer surface area covered. Several authors have mistakenly tried to schedule controller tuning based on liquid level for reactor temperature control. One author has reported being bewildered by its failure. This is not the case for gas pressure control. The equations show that liquid level has a profound effect on the process integrating gain for vessel pressure control because it changes the vapor space volume without any competing effect. To summarize, the integrator gain for composition and gas pressure is inversely proportional to liquid level (liquid mass). For temperature, the effect of level cancels out unless the level is above or below the heat transfer surface area, which is unusual but can occur at the beginning or end of a batch when coils instead of a jacket is used for heat transfer. For temperature, the integrator gain is nearly always proportional to the overall heat transfer coefficient that is a function of mixing, process composition, and fouling or frosting.

The equations also show that if the transport delay for flow injection is large compared to the time constant, which does occur for reagent injection in dip tubes for pH control), then the controller gain will be proportional to flow. Note that pH control is a class of concentration control.

For the control of temperature and concentration in a pipe, the process dead time and process gain are both inversely proportional to flow and the process time constant is essentially zero, which makes the actuator, sensor, transmitter, or signal filter time lag the largest time constant in the loop. Thus, the largest automation system lag determines the dead time to time constant ratio. For a static mixer, there is some mixing, and the process time constant is inversely proportional to flow but is usually quite small compared to other lags in the loop. The controller gain is generally proportional to flow for both cases.

Finally, the above has implications so far as whether a flow feedforward multiplier or summer and whether a linear or equal percentage trim should be used. A flow feedforward multiplier and equal percentage trim, which both have a gain proportional to flow, can help compensate for a process gain that is inversely proportional to flow provided the process time constant is not also inversely proportional to flow. This is generally the case for temperature and concentration control of plug flow volumes (pipelines, static mixers, and heat exchangers). For well-mixed volumes, feedforward summers and an installed linear characteristic for valves is generally best. For control valves this corresponds to a linear trim when the available pressure drop that is much larger than the system pressure drop or critical pressure drop so the installed is close to the inherent flow characteristic.

The results are also useful for determining the dead time to time constant ratio, which has a profound effect on the tuning factors used and the performance of dead time compensation, which has been discussed on the *Control* magazine voices Control Talk Blog site.

Batch Implications

Plug flow volumes can always be considered as continuous because the volume is completely full and anything entering will be discharged after a transportation delay.

A back mixed volume is partially full. If the liquid discharge flow is zero, this volume can be considered to be in the batch mode rather than the continuous mode. If the flows are all sequenced and charged based on time and/or totals, the vessel operation can be considered to be pure batch. If reactor flows are ratioed and manipulated by a control loop, the vessel operation can be classified as fed-batch.

Level has an integrating response whether in the batch or continuous mode. In the batch mode, it is a zero load integrator in that all the feed flows must be zero for the level to stop rising. Level has a one sided integrating response in the batch mode since the level can only rise and not drop. This type of response causes overshoot for any controller with reset action. Proportional plus derivative (PD) controllers with a zero or negative bias can be tuned for zero overshoot.

The temperature response of a back mixed volume remains self-regulating even for zero liquid discharge flow unless the liquid level is above or below the heat transfer surfaces. However, the temperature response does lose some self-regulation and behaves more like a "near integrator".

The concentration response of a back mixed volume becomes integrating for a zero discharge flow. This is not obvious because the discharge flow cancels out of the differential equation from the application of the multiplicative rule of integration in the transition of the derivative from the rate of accumulation of component mass to the rate of accumulation of component concentration. The effect of zero discharge flow is more recognizable if we consider the case of a zero reaction rate so that the process time constant is simply the residence time (liquid mass divided by the liquid feed flow rate). The increase in mass for a fixed feed rate over the residence time is simply the feed rate multiplied by the residence time. The result is an increase in mass equal to the existing mass. This doubling of mass doubles the residence time and hence the process time constant. Consequently, the process never reaches a steady state because the process time constant is constantly increasing as the level is rising for a zero discharge flow. For the case of zero reaction, the integrator is a zero load integrator because the feed of the component must be zero for the concentration of the component to stop rising. The concentration here has a one sided integrating response in that the concentration can only increase and not decrease. This would also be the case for reaction products where there is only a forward reaction (no reverse or side reactions). As with level, overshoot is a problem unless PD controllers are used. Alternately, the controlled variable can be translated to a rate of change of concentration as noted in application of model predictive control for bioreactor biomass and product concentration.

Gas pressure is an integrator regardless of liquid discharge flow as long as the pressure in the vessel has a negligible effect on vent flow, which is the case for large or critical pressure drops. If this is not the case, the gas pressure response becomes self-regulating but for large volumes and small vent flows it behaves like a "near integrator."

Results

The integrating process gain (K_{ip}) for the control of liquid level by the manipulation of a flow:

$$K_{ip} = 1/(\rho_o * A_o) \quad \text{(F.4d)}$$

The integrating process gain (K_{pi}) for the control of pressure per gas law by the manipulation of a flow:

$$K_{ip} = [(R_g * T_g)/V_g] \quad \text{(F.5d)}$$

For the manipulation of jacket temperature to control outlet temperature, the main process time constant (τ_p) is (positive feedback if heat of feed and reaction exceeds product of heat transfer coefficient and area):

$$\tau_p = (C_p * M_o)/[C_p * F_f - \Delta Q_r/\Delta T_o + U * A] \quad \text{(F.6g)}$$

For the manipulation of jacket temperature to control outlet temperature, the integrating process gain (K_p) is:

$$K_p = (U * A)/[C_p * F_f - \Delta Q_r/\Delta T_o + U * A] \quad \text{(F.6h)}$$

For the manipulation of jacket temperature to control outlet temperature, the near-integrating process gain (K_{nip}) is:

$$K_{nip} = (U * A)/(C_p * M_o) \quad \text{(F.6i)}$$

For the manipulation of feed temperature to control outlet temperature, the process gain (K_p) is:

$$K_p = (C_p * F_f)/[C_p * F_f - \Delta Q_r/\Delta T_o + U * A] \quad \text{(F.6j)}$$

For the manipulation of feed flow to control outlet temperature, the process gain (K_p) is:

$$K_p = (C_p * T_f)/[C_p * F_f - \Delta Q_r/\Delta T_o + U * A] \quad \text{(F.6k)}$$

For manipulation of jacket temperature, the additional small secondary process time constant associated with the heat capacity and mass of the jacket wall is:

$$\tau_{p2} = (C_w * M_w)/[U * A] \quad \text{(F.6l)}$$

The process dead time (θ_p) from the turnover time for temperature and concentration control in a well mixed volume is:

$$\theta_p = (M_o/\rho_o)/[(F_f + F_a + F_r)/\rho_o + F_v/\rho_v] \quad \text{(F.6m)}$$

The process dead time (θ_p) from injection delay for concentration control is:

$$\theta_p = V_1/(F_1/\rho_1) \quad \text{(F.6n)}$$

For the manipulation of feed flow to control reactant concentration (X_{Ao}), the main process time constant (τ_p) is:

$$\tau_p = M_o/(R_x + F_f) \quad \text{(F.7g)}$$

For the manipulation of feed flow to control reactant concentration (X_{Ao}), the process gain (K_p) is:

$$K_p = X_{Af}/(R_x + F_f) \quad \text{(F.7h)}$$

For the manipulation of feed flow to control reactant concentration, the near-integrating process gain (K_{nip}) is:

$$K_{nip} = X_{Af}/M_o \tag{F.7i}$$

For the manipulation of feed concentration to control reactant concentration, the process gain (K_p) is:

$$K_p = F_f/(R_x + F_f) \tag{F.7j}$$

For plug flow volumes where different streams are being combined, the process gain (K_p) for controlling the temperature of the mixture (T_f) by the manipulation of flow is:

$$K_p = dT_f/dF_1 = T_1/\sum F_i \tag{F.9a}$$

For plug flow volumes where different streams are being combined, the process gain (K_p) for controlling the composition of component A (X_{Af}) in the mixture by the manipulation of flow is:

$$K_p = dX_{Af}/dF_1 = X_{A1}/\sum F_i \tag{F.9b}$$

For plug flow volumes where different streams are being combined, the process dead time for controlling the temperature or composition of the mixture by the manipulation of flow is:

$$\theta_o = V_1/(F_1/\rho_1) + V_p/\sum(F_i/\rho_i) \tag{F.9c}$$

The process time constant is essentially zero for true plug flow. For a static mixer there is some back mixing, the residence time in Equation F.9c is split between a dead time and time constant per equations F.9d and F.9e.

$$\theta_p = V_1/(F_1/\rho_1) + x * V_p/\sum(F_i/\rho_i) \tag{F.9d}$$

$$\tau_p = (1 - x) * V_p/\sum(F_i/\rho_i) \tag{F.9e}$$

Derivations

There are three types of processes; self-regulating, integrating, and runaway as shown in Figures F.1, F.2, and F.3, respectively. A self-regulating process will decelerate to a new steady state. An integrating process will continually ramp. A runaway process will accelerate until hitting a relief or interlock setting.

Over 90% of the processes are self-regulating. However, many of the continuous and fed-batch processes in the chemical industry with the greatest direct economic benefits behave and can be best treated as "near-integrating" processes. The classic integrating process is a pure batch or level process. Less than 1% of the processes are runaway. When these exist, understanding the runaway response is critical in terms of safety and control because of the propensity to accelerate and reach a point of no return. Runaway responses are almost exclusively associated with highly exothermic reactors used in plastics and specialty chemical production.

This note develops the equations for process dynamics for back mixed volumes and plug flow volumes. The back mixed volumes section applies to volumes whenever an agitator pumping rate, an eductor or recirculation liquid flow rate, or gas evolution or sparge rate produces enough turbulence and back mixing to make the mixture more uniform in the axial besides the radial direction of the volume. Gas volumes generally have enough turbulence and a fast enough gas dispersion rates to be treated as a back mixed volume. The equations therefore hold relatively well for an evaporator and a single distillation stage due to turbulence from the vapor flow. The plug volume section applies to static mixers, pipelines, coil inlets, and jacket inlets where the turbulence from pipe fittings or internal mixing elements creates enough radial mixing to make the mixture uniform over the cross section of the pipe or nozzle inlet but little axial mixing.

Back Mixed Volumes

For a back mixed volume, the process gains and time constants can be readily identified if the ordinary differential equations for the rate of accumulation of energy or material in the volume are set up so that the process output of interest (Y) is on right side with a unity coefficient. From this simple generic form we can identify the process time constant (τ_p) as the coefficient of derivative of the process output (dY/dt) and the process gain (K_p) as the coefficient of the

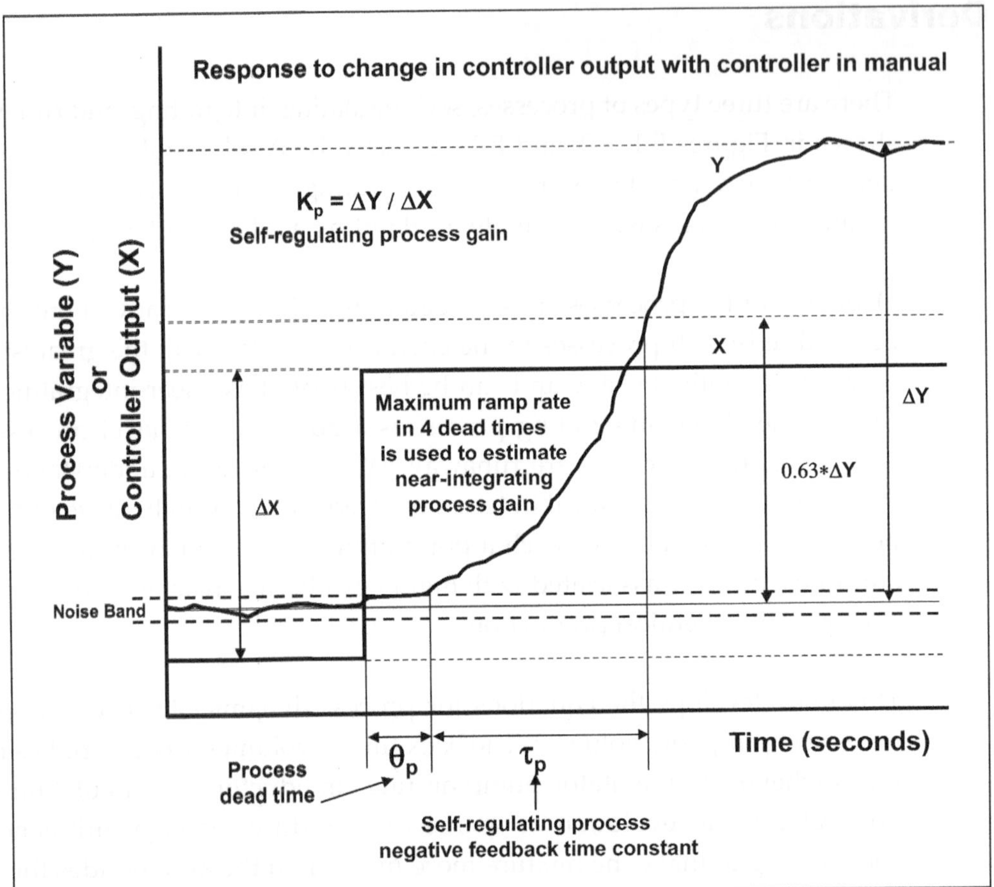

Figure F.1 – *Self-Regulating* (Negative Feedback) Process

process input (X). The process output (Y) and input (X) can be viewed as the controlled and manipulated variables, respectively. Many other terms can exist but these are not shown in the following equations. These missing terms can be categorized as disturbances.

If the sign of the unity coefficient of the process output on the right side is negative (Equation F.1a), the process has negative feedback. As the process output changes, the negative feedback slows down and eventually halts the excursion of the process output at its new steady state when it balances out the effect of the process input and the disturbances.

$$\tau_p * dY/dt = K_p * X - Y \tag{F.1a}$$

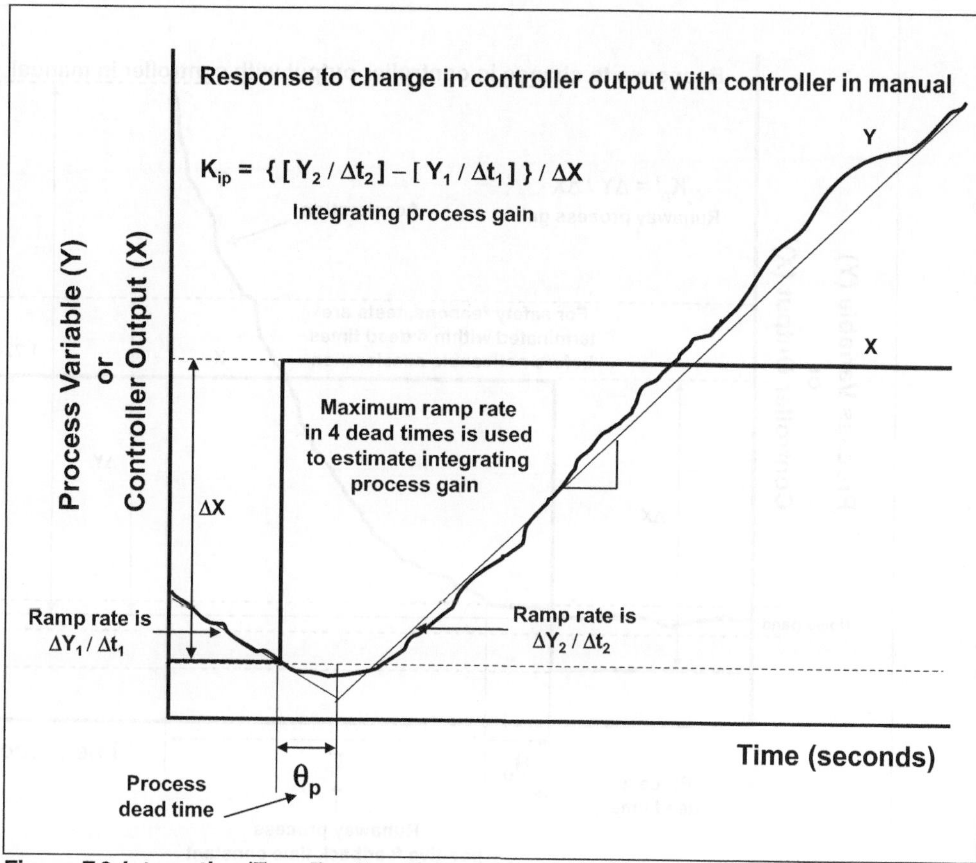

Figure F.2 *Integrating* (Zero Feedback) Process

The integration of this equation provides the time response of a change in the process output (ΔY) for a step change in the process input (ΔX). The step occurs at $t = 0$.

$$\Delta Y = K_p * (1 - e^{-t/\tau_o}) * \Delta X \tag{F.1b}$$

If the process output does not appear on the right side (Equation F.2a), there is no process feedback. As the process output changes, there is no feedback to slow it down or speed it up so it continues to ramp. There is no steady state. The ramping will only stop when X is zero or balances out the disturbances.

$$dY/dt = K_i * X \tag{F.2a}$$

$$\Delta Y = K_i * t * \Delta X \tag{F.2b}$$

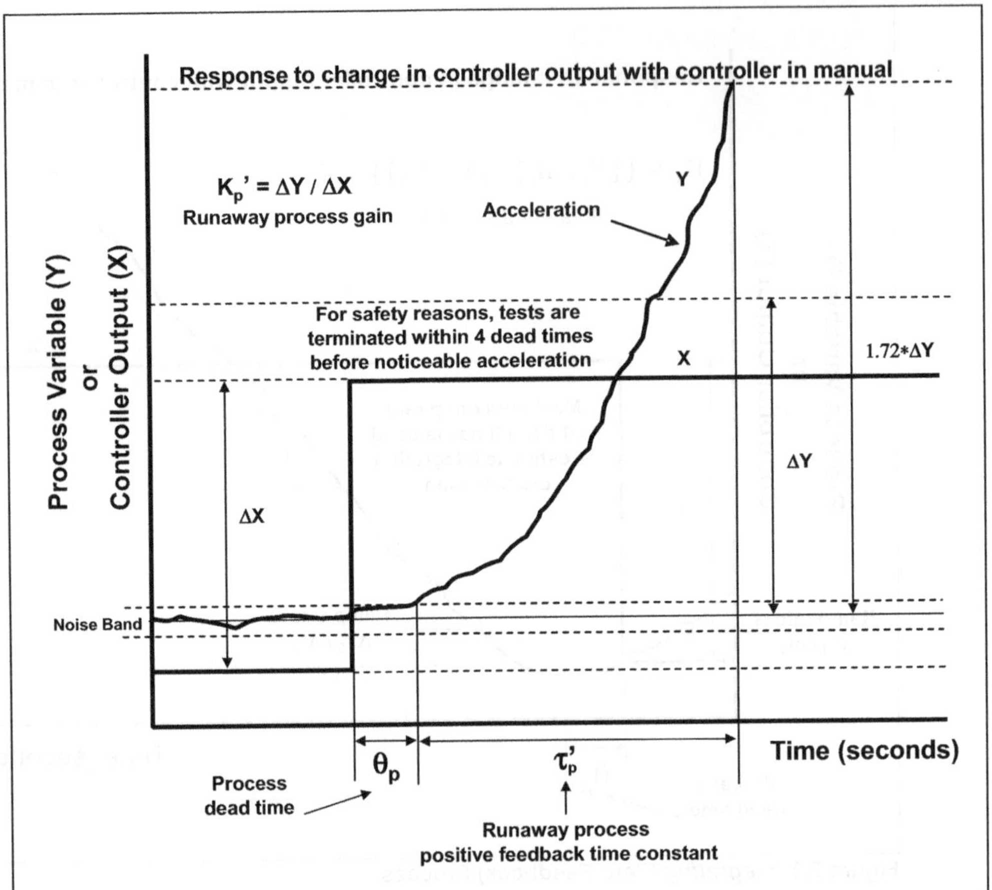

Figure F.3 ***Runaway*** **(Positive Feedback) Process**

Often in the more important loops for concentration, pressure, and temperature control of large volumes, the time constant in equation F.1a is so large that the time to reach steady state is beyond the time frame of interest. Since these loops with small dead time to time constant ratios should be tuned with small lambda factors (high controller gains) per Advanced Application Note 3, the controller only sees the first part of the excursion before the inflection point and deceleration by negative process feedback. In this case the response is best characterized by a near-integrating process gain calculated per Equation F.2c.

$$K_{nip} = K_p / \tau_p \qquad \text{(F.2c)}$$

If the sign of the unity coefficient of the process output on the right side is positive (Equation F.3a), the process has positive feedback. As the process output changes, the positive feedback speeds up the excursion unless disturbances counteract the effect of the process input and output.

$$\tau_p' * dY/dt = K_p * X + Y \tag{F.3a}$$

$$\Delta Y = K_p * (e^{t/\tau_o'} - 1) * \Delta X \tag{F.3j}$$

Consider a mixed volume with a jacket and vapor space as shown in Figure F.4. There are liquid reactant feeds, gas feeds (sparged through the liquid and added directly to the vapor space), an outlet liquid flow, a vent gas flow, and a jacket coolant flow. There is normally multiple components interest. For example, consider liquid or gas acid and base reagent or reactant components (a, b) to produce primary and secondary liquid or gas products (c, d, e). Consider also there are typically water and nitrogen gas components (w, n).

The ordinary differential equation for the accumulation of liquid mass as shown in Equation F.4a includes inlet flows added directly to the liquid volume (ΣF_i), vapor flow rates from evaporation and vaporization (ΣF_v), and an outlet liquid flow rate (F_o). The liquid level depends upon density and cross section area of the liquid. Equation F.4a can then be reformulated to Equation F.4b to include the process variable of interest, liquid level (L_i), in the derivative.

$$dM_o/dt = \sum F_i - \sum F_v - F_o \tag{F.4a}$$

$$d(\rho_o * A_o * L_o)/dt = \sum F_i - \sum F_v - F_o \tag{F.4b}$$

If we consider the density (ρ_o) to be a weak function of composition and therefore constant like the cross sectional area (A_o) we can take these terms outside the derivative and divide through to get an equation for level (L_o) in the form of F.2a. Now it is clearly evident that the integrating process gain (K_{ip}) for manipulation of flows in or flow out is simply the inverse of the product of the liquid density and cross section area (Equation F.4d).

$$dL_o/dt = [1/(\rho_o * A_o)] * \left[\sum F_i - \sum F_v - F_o\right] \tag{F.4c}$$

$$K_{ip} = 1/(\rho_o * A_o) \tag{F.4d}$$

The ordinary differential equation for the accumulation of gas mass as shown in Equation F.5a includes inlet flows added directly to the gas volume (ΣF_i), vapor flow rates from gas sparging, evolution, and vaporization (ΣF_v), and an exit gas flow rate (F_g). Equations of state such as the ideal gas law can be used to express this relationship for a given composition. Equation F.5a can then be reformulated to Equation F.5b to include the process variable of interest, gas pressure (P_g), in the derivative.

$$dM_g/dt = \sum F_i + \sum F_v - F_g \quad \text{(F.5a)}$$

$$d[(P_g * V_g)/(R_g * T_g)]/dt = \sum F_i + \sum F_v - F_g \quad \text{(F.5b)}$$

If we consider changes in the gas volume (V_g) and gas temperature (T_g) to be much slower than changes in the gas pressure (P_g) and therefore relatively constant during the integration step we can take these terms outside the derivative and divide through to get an equation for pressure in the form of Equation F.2a. Now it is clearly evident that the integrating process gain (K_{ip}) for manipulation of flows in or flow out is simply the product of the universal gas coefficient (R_g) and the absolute gas temperature divided by the gas volume (Equation F.5d). This assumes a change in pressure does not significantly change the gas glow out of the volume, which is normally the case for a pressure drop across the vent valve that is large or critical.

$$dP_g/dt = [(R_g * T_g)/V_g] * \left\lfloor \sum F_i + \sum F_v - F_g \right\rfloor \quad \text{(F.5c)}$$

$$K_{ip} = [(R_g * T_g)/V_g] \quad \text{(F.5d)}$$

The ordinary differential equation (ODE) for the accumulation of energy as shown in Equation F.6a includes the effects of feed temperature, heat of reaction as a function of temperature, heat of vaporization, and heat transfer to the jacket. If we consider the specific heat capacity relatively constant and use the multiplicative rule of integration, we can express the differential equation in the generic form of Equation F.1a in terms of temperature to show the process feedback in Equation F.6f, the final form of the ODE. The relative magnitude of the terms in the denominator of Equation F.6g determines the feedback sign.

$$dQ_o/dt = C_p{*}\sum(F_i{*}T_i) - C_p{*}F_o{*}T_o + (\Delta Q_r/\Delta T_o){*}T_o \tag{F.6a}$$
$$+ H_x{*}R_x - H_v{*}F_v - U{*}A{*}(T_o - T_j)$$

$$dQ_o/dt = d(C_p{*}M_o{*}T_o)/dt = C_p{*}(dM_o/dt){*}T_o + C_p{*}M_o{*}(dT_o/dt) \tag{F.6b}$$

$$F_f = \sum F_i \tag{F.6c}$$

$$T_f = \sum(F_i * T_i)/\sum F_i \tag{F.6d}$$

$$C_p{*}(dM_o/dt){*}T_o = C_p{*}(F_f - F_o){*}T_o \tag{F.6e}$$

$$C_p * M_o * (dT_o/dt) = C_p * F_f * T_f + H_x * R_x - H_v * F_v + U * A * T_j \tag{F.6f}$$
$$-[C_p * F_f - \Delta Q_r/\Delta T_o + U * A] * T_o$$

For the manipulation of jacket temperature to control outlet temperature, the main process time constant (τ_p) is (positive feedback if heat of feed and reaction exceeds product of heat transfer coefficient and area):

$$\tau_p = (C_p * M_o)/[C_p * F_f - \Delta Q_r/\Delta T_o + U * A] \tag{F.6g}$$

For the manipulation of jacket temperature to control outlet temperature, the process gain (K_p) is:

$$K_p = (U * A)/[C_p * F_f - \Delta Q_r/\Delta T_o + U * A] \tag{F.6h}$$

For the manipulation of jacket temperature to control outlet temperature, the near-integrating process gain (K_{nip}) is:

$$K_{nip} = (U * A)/(C_p * M_o) \tag{F.6i}$$

For the manipulation of feed temperature to control outlet temperature, the process gain (K_p) is:

$$K_p = (C_p * F_f)/[C_p * F_f - \Delta Q_r/\Delta T_o + U * A] \tag{F.6j}$$

For the manipulation of feed flow to control outlet temperature, the process gain (K_p) is:

$$K_p = (C_p * T_f)/[C_p * F_f - \Delta Q_r/\Delta T_o + U * A] \quad \text{(F.6k)}$$

For manipulation of jacket temperature, the additional small secondary process time constant associated with the heat capacity and mass of the jacket wall is:

$$\tau_p = (C_w * M_w)/[U * A] \quad \text{(F.6l)}$$

Any change in the temperature at the heat transfer surfaces or the feed inlet must be dispersed and back mixed into the volume. This process dead time (θ_p) is the turn over time that can be approximated as the liquid inventory divided by the summation of the feed flow rate (F_f), agitator pumping rate (F_a), recirculation flow rate (F_r), and vapor evolution rate or vapor bubble rate (F_v). Since this turn over time is computed in terms of volumetric flow rates, the liquid mass and the mass flow rates are divided by their respective densities as shown in Equation F.6m.

$$\theta_p = (M_o/\rho_o)/[(F_f + F_a + F_r)/\rho_o + F_v/\rho_v] \quad \text{(F.6m)}$$

If there is an injector (dip tube or sparger ring) volume, a change in composition at the nozzle must propagate by plug flow to the discharge points of the dip tube or sparger ring. The dead time for a feed flow (F_1) is the injector volume (V_1) divided by the injection mass flow (F_1) divided by its respective density (ρ_1).

$$\theta_p = V_1/(F_1/\rho_1) \quad \text{(F.6n)}$$

The ordinary differential equation (ODE) for the accumulation of liquid reactant mass (M_A) as shown in Equation F.7a includes the effects of feeds (F_i) with a reactant mass fraction (X_{Ai}), reaction rate (R_x), and outlet flow (F_o). The feeds can be from raw material, intermediate products, recycle streams, or multi-stage reactors. If we use the multiplicative rule of integration, we can express the differential equation in the generic form of Equation F.1a in terms of concentration to show the process feedback in Equation F.7f, the final form of the ODE.

$$dM_A/dt = \sum(F_i * X_{Ai}) - (R_x + F_o) * X_{Ao} \quad \text{(F.7a)}$$

$$dM_A/dt = d(M_o * X_{Ao})/dt = (dM_o/dt) * X_{Ao} + M_o * (dX_{Ao}/dt) \quad \text{(F.7b)}$$

$$F_f = \sum F_i \quad \text{(F.7c)}$$

$$X_{Af} = \sum(F_i * X_{Ai}) / \left(\sum F_i\right) \quad \text{(F.7d)}$$

$$(dM_o/dt) * X_{Ao} = (F_f - F_o) * X_{Ao} \quad \text{(F.7e)}$$

$$M_o * (dX_{Ao}/dt) = F_f * X_{Af} - (R_x + F_f) * X_{Ao} \quad \text{(F.7f)}$$

For the manipulation of feed flow to control reactant concentration (X_{Ao}), the main process time constant (τ_p) is:

$$\tau_p = M_o/(R_x + F_f) \quad \text{(F.7g)}$$

For the manipulation of feed flow to control reactant concentration (X_{Ao}), the process gain (K_p) is:

$$K_p = X_{Af}/(R_x + F_f) \quad \text{(F.7h)}$$

For the manipulation of feed flow to control reactant concentration, the near integrator gain (K_i) is:

$$K_i = X_{Af}/M_o \quad \text{(F.7i)}$$

For the manipulation of feed concentration to control reactant concentration, the process gain (K_p) is:

$$K_p = F_f/(R_x + F_f) \quad \text{(F.7j)}$$

The process dead times from turnover time and from feed injection are the same as computed in the section for temperature control (Equations F.6m and F.6n).

Plug Flow Volumes

For plug flow volumes where different streams are being combined, the process gain for controlling the temperature (T_f) or composition (X_{Af}) of the mixture (often a feed to a downstream equipment) can be computed by taking the derivative of Equations F.6d and F.7d with respect to the manipulated flow stream 1 (F_1) to give Equations F.9a and F.9b, respectively. In both cases, the process gain is inversely proportional to total flow (ΣF_i).

$$K_p = dT_f/dF_1 = T_1/\sum F_i \qquad \text{(F.9a)}$$

$$K_p = dX_{Af}/dF_1 = X_{A1}/\sum F_i \qquad \text{(F.9b)}$$

The process dead time for the manipulation of a flow for stream 1 (F_1) is the summation of the injection delay for steam 1 and the piping delay from the point of injection to the point of temperature or composition measurement. For plug flow the residence time, which is the second expression in Equation F.9c completely becomes dead time.

$$\theta_p = V_1/(F_1/\rho_1) + V_p/\sum(F_i/\rho_i) \qquad \text{(F.9c)}$$

The process time constant is essentially zero for true plug flow. For a static mixer there is some back mixing, the residence time in Equation F.9c is split between a dead time and time constant per equations F.9d and F.9e.

$$\theta_p = V_1/(F_1/\rho_1) + x * V_p/\sum(F_i/\rho_i) \qquad \text{(F.9d)}$$

$$\tau_p = (1-x) * V_p/\sum(F_i/\rho_i) \qquad \text{(F.9e)}$$

It is obvious from the above that both the process gain and dead time are inversely proportional to total flow.

Controller Tuning

The implication of the results can be best seen if lambda is set equal to the total loop dead time (θ_o) resulting in Equation F.10a for the PID gain. If the open loop time constant (τ_o) is large compared to the dead time, the process response is termed near-integrating and the ratio of the open loop self-regulating process gain (K_o) to the time constant is used to approximate the open loop integrating process gain (K_i) in Equation 5.10b. This PID gain is about ½ of the PID gain estimated by the Ziegler Nichols reaction curve method.

$$K_c = 0.5 * \frac{\tau_o}{K_o * \theta_o} \tag{F.10a}$$

$$K_c = 0.5 * \frac{1}{K_i * \theta_o} \tag{F.10b}$$

The open loop time constant (τ_o) in the numerator of Equation F.10a is the largest time constant in the loop wherever it occurs. Hopefully, the process is mixed well enough and the instrumentation is fast enough that the largest time constant is in the process ($\tau_o = \tau_p$) and not the automation system. A large time constant in the process slows down the disturbance and is desirable. A large time constant in the measurement and final element is detrimental because it slows down the ability of the controller to see and react to disturbance, respectively.

The open loop self-regulating process gain (K_o) in the denominator is dimensionless. The process gain is actually the product of the manipulated variable gain, the process gain (K_p), the gain of nonlinear process variables, such as pH (slope of the titration curve), and the controlled variable gain. For a loop that throttles a control valve, the manipulated variable gain is the slope of the valve's installed characteristic. For the primary loop of a cascade control system, the manipulated variable gain is the secondary loop setpoint span divided by 100%. The controlled variable gain is 100% divided by the process variable span. Thus, changes in calibration span affect the computed controller gain, which provides robustness and a less oscillatory response.

Finally, the dead time (θ_o) in the denominator is really the total loop dead time, which is summation of the process dead time (θ_p) plus all the small time lags and delays in the loop. While the names open loop time constant (τ_o), open loop self-regulating process gain (K_o), and total loop dead time (θ_o) for the parameters in Equation F.10a are more definitive, nearly all of the control literature uses the terms process time constant, process gain, and process dead time indiscriminately.

Nomenclature

Process Parameters:

A_o = cross sectional area of liquid level (m2)
A = heat transfer surface area (m^2)
C_p = heat capacity of process (kJ/kg*°C)
C_w = heat capacity of wall of heat transfer surface (kJ/kg*°C)
F_a = agitator pumping rate (kg/sec)
F_f = total feed flow (kg/sec)
F_g = gas flow (kg/sec)
F_i = feed stream i flow (kg/sec)
F_o = vessel outlet flow (kg/sec)
F_r = recirculation flow (kg/sec)
F_v = vaporization rate (kg/sec)
H_v = heat of vaporization (kJ/kg)
H_x = heat of reaction (kJ/kg)
L_o = liquid level (m)
M_A = component A mass (kg)
M_g = gas mass (kg)
M_o = liquid mass (kg)
M_w = mass of wall of heat transfer surface (kg)
P_g = gas pressure (kPa)
T_f = total feed temperature (°C)
T_g = gas temperature (°C)
T_i = feed stream i temperature (°C)
T_o = vessel outlet temperature (°C)
t = time (sec)
Q_o = total heat of liquid (kJ)
Q_r = heat from reaction (kJ)
R_x = reaction rate (kg/sec)
R_g = universal constant for ideal gas law (kPa/m^3/°C)
ρ_g = gas density (kg/m^3)
ρ_i = stream i density (kg/m^3)
ρ_o = liquid density (kg/m^3)
ρ_v = density of vapor (kg/m^3)
U = overall heat transfer coefficient (kJ/m^2*°C)
V_g = gas volume (m^3)
V_i = injection (e.g., dip tube or sparger ring) volume (m^3)
V_p = piping volume (m^3)
x = fraction of volume that is plug flow

X_{Af} = total feed component A concentration (mass fraction)
X_{Ai} = feed stream i component A concentration (mass fraction)
X_{Ao} = vessel outlet component A concentration (mass fraction)

Generic Terms:

X = process input (manipulated variable) (eu)
Y = process output (controlled variable) (eu)

Dynamic Parameters:

K_c = PID controller gain (dimensionless)
K_i = open loop integrating process gain (1/sec)
K_{ip} = integrating process gain (eu/eu)
K_o = open loop self-regulating process gain (dimensionless)
K_p = process gain (eu/eu)
K_{nip} = near-integrating process gain (eu/eu/sec)
τ_p = negative feedback process time constant (sec)
τ_p' = positive feedback process time constant (sec)
τ_o = open loop time constant (sec)
θ_p = process dead time (sec)
θ_o = total loop dead time (sec)

References

1. Al-Khalifah, M. and McMillan, G. "Control Valves versus Variable Speed Drive for Flow Control" *ISA Automation Week*, Orlando, 2012.

2. Arant, J. "Control Valve Notes and Positioner Myths - Part I" *Control*, Nov, 1993.

3. Arant, J. "Control Valve Notes and Positioner Myths - Part II" *Control*, Dec, 1993.

4. Åström, K. and Hägglund, T. *Advanced PID Control*, ISA, 2006.

5. Bialkowski, W. "Process Control Related Variability and the Link to End-Use Performance" *Control Systems 90*, Helsinki, Finland, 1990.

6. Bialkowski, W. and Janvier, B. "On Lambda Tuning – How and Why" *PAPTAC 91st Annual Meeting*, 2005.

7. Bailey, J. and Ollis, D. *Biochemical Engineering Fundamentals* McGraw-Hill, New York, 1977.

8. Bergeron, L. *Water Hammer in Hydraulics* Wiley, New York, 1961.

9. Blevins, T. et al., *Advanced Control Unleashed*, ISA, 2003.

10. Bohl, A. and McAvoy, T. "Linear Feedback vs. Time Optimal Control II - The Regulator Problem" *Ind. Eng. Chem.*, Process Des. Dev., Vol. 15, No. 1, 1976.

11. Bor, T. "The Static Mixer as a Chemical Reactor" *British Chemical Engineering*, Vol. 16, No. 7, 1971.

12. Boudreau, M. and McMillan, G. *New Directions in Bioprocess Modeling and Control*, ISA, 2006.

13. Bradner, M. "Pneumatic Transmission Lag" *Instruments*, Vol. 22, July 1949.

14. Brez, C. "Predicting Limit Cycling in Control Valves" *Control Engineering*, Vol. 22, No. 7, 1975.

15. Buchwald, W. "Control Performance with Reduced Valve Pressure Drop for Energy Conservation" Advances in Instrumentation, *ISA International Conference*, New York, Vol. 29, Part 2, Paper 610, October 1974.

16. Buckley, P. "Dynamics of Temperature Measurement" *Process Annual Symposium*, Instrumentation Process Industry, Wilmington, Vol. 34, 1979.

17. Coughran, M. "Valves: Testing for Peak Performance", *InTech*, October 1994.

18. Coughran, M. "Performance Influences in Globe Control Valves", *InTech*, August 1996.

19. Coughran, M. "Lambda Tuning—the Universal Method for PID Controllers in Process Control" White Paper, *Control*, 2014.

20. Fitzgerald, Bill, *Control Valves for the Chemical Process Industries*, McGraw-Hill, Inc., 1995.

21. Giusti, A. and Hougen, J. "Dynamics of pH Electrodes" *Control Engineering*, Vol. 8, No. 4, 1961.

22. Green, R. and Field, W. "Evaluation of pH Measurement Response" *AID Symposium Proceedings*, San Francisco, June 1964.

23. Hägglund, T. "A Control-Loop Performance Monitor" *Control Eng. Practice* Vol. 3, No. 11, *Elsevier Science Ltd*, 1995.

24. Harriott, P. *Process Control* McGraw-Hill, 1964.

25. Heider, R. "How Digital Controllers Can Cause Process Cycling" *ACC Conference Proceedings*, Arlington, Va., June 1982.

26. Heider, R. and Clay, L. "Make the Most of Variable Frequency Drives", *Control*, January 2009.

27. Hopkins, B. and Alford, G. "Temperature Control of Polymerization Reactor", *Instrumentation Technology*, May 1993.

28. Kardos, P. "Response of Temperature Measuring Elements", *Chemical Engineering*, Aug. 29, 1977.

29. Keith, G. "Linear Installed Characteristic Simplify Control Valve Selection", *Control Engineering*, November 1977.

30. Kerlin, T., et al., "Time Response of Installed Temperature Sensors" *Sixth International Symposium on Temperature*, Paper 19-2, 1982.

31. Latour, L., et al. "Estimating Benefits from Advanced Control", *ISA Transactions*, Vol. 25, No.4, 1986.

32. Lipták, B. *Instrument Engineers Handbook* Chilton, 2010.

33. Luyben, W. and Cheung, T. "Proportional-Derivative Control of Exothermic Reactors" *Joint Automatic Control Conference*, Philadelphia, Vol. 3, October, 1978.

34. Luyben, W. and Melcic, M. "Consider Reactor Control Lags", *Hydrocarbon Processing*, Vol. 57, No. 3, 1978.

35. Luyben, W. and Cheng-Ching, Y. "Analysis of Valve-Position Control for Dual Input Processes", *Ind. Eng. Chem. Fundam*, Vol.25, No. 3, 1986.

36. Luyben, W. and Tyreus, B. "Dynamics and Control of Recycle Systems. 4. Ternary Systems with One or Two Recycle Streams" *Ind. Eng. Chem. Res*, Vol. 32, No. 6, 1993.

37. Luyben, W. and Luyben, M. *Essentials of Process Control* McGraw-Hill, 1997.

38. Luyben, W. *Chemical Reactor Design and Control*, Wiley, 2007.

39. Mamzic, C. "Improving the Dynamics of Pneumatic Positioners", *ISA Journal*, Vol. 5, No. 8, August 1958.

40. McMillan, G. "The Effect of Process Non-Self-Regulation on Control Loop Performance", *ISA Conference Proceedings*, Anaheim, Oct. 1981.

41. McMillan, G. "The Effect of Cascade Control on Loop Performance", *ACC Conference Proceedings*, Arlington, June 1982.

42. McMillan, G. *Process/Industrial Instruments and Controls Handbook 5th Edition*, McGraw-Hill, 1999.

43. McMillan, G. "What's Your Flow Control Valve Telling You?", *Control Design*, May 2004.

44. McMillan, G. *Advanced pH Measurement and Control, Third Edition*, ISA, 2004.

45. McMillan, G. "Full Throttle Batch and Startup Response", *Control*, May 2006.

46. McMillan, G. et al. "Maximizing Process Benefits form Bioprocess Modeling and Control", *Pharmaceutical Technology*, Nov 2006.

47. McMillan, G. and Sowell, M. "Virtual Control of Real pH", *Control*, November 2007

48. McMillan, G. and Boudreau, M. *New Directions in Bioprocess Modeling and Control* ISA, 2007.

49. McMillan, G. et al. "PAT Tools for Accelerated Process Development and Improvement" Process Design Supplement, *BioProcess International*, March 2008.

50. McMillan, G., Reedy, R. and Moulis, J. "Virtual Plant Provides Real Insights" *Chemical Processing*, December 2009.

51. McMillan, G. *Essentials of Modern Measurements and Final Elements in the Process Industries*, ISA, 2010.

52. McMillan, G. *Centrifugal and Axial Compressor Control*, Momentum Press, 2010.

53. McMillan, G. *Process Control Case Histories*, Momentum Press, 2010.

54. McMillan, G. "Wireless Measurement and Control Opportunities", *ISA Automation Week*, Mobile, 2011.

55. McMillan, G. "pH Measurement and Control Opportunities", *ISA Automation Week*, Mobile, 2011.

56. McMillan, G. "Reactor Control Opportunities", *ISA Automation Week*, Mobile, 2011.

57. McMillan, G. "Corn Analysis Modeling and Control for Ethanol Yield Improvement", *ISA Automation Week*, Mobile, 2011.

58. McMillan, G. "Don't Overlook PID in APC", *Control*, November 2011.

59. McMillan, G. "Advances in Flow and Level Measurements Enhance Process Knowledge, Control", *InTech*, January/February 2010.

60. McMillan, G. "Unlocking the Secret Profiles of Batch Reactors" *Control*, Jul. 2008.

61. McMillan, G. "Is Wireless Process Control Ready for Prime Time" *Control*, May 2009.

62. McMillan, G. "Wireless - Overcoming Challenges of PID Control, Analyzer Applications" *InTech*, July/August 2010.

63. McMillan, G. and Baril, R. "pH Measurement and Control", *Chemical Engineering*, August 2010.

64. McMillan, G. "Temperature Measurement, Control Key to Plant Performance", *InTech*, November/December 2010.

65. McMillan, G. "Feedforward Control Enables Flexible, Sustainable Manufacturing", *InTech*, March/April 2011.

66. McMillan, G. *Advanced Temperature Measurement and Control, Second Edition*, ISA, 2011.

67. McMillan, G. "Get the Most Out of Your Batch", *Control*, September, 2012.

68. McMillan, G. "PID Tuning Rules", *InTech*, January/February 2012.

69. McMillan G. and Torres, H. "Effective Use of Key PID Features", ISA Automation Week, Orlando, 2012.

70. McMillan, G. *Good Tuning: A Pocket Guide, Third Edition*, ISA, 2012.

71. McMillan, G. and Vegas, H. *101 Tips for a Successful Automation* Career, ISA, 2012.

72. McMillan, G. "Maximize pH Response, Accuracy and Reliability", *Control*, February 2013.

73. Moore, R. "Good pH Measurements in Bad Process Streams", *Instruments and Control Systems*, December 1990.

74. Nelson, D. "Better Damper Control Nixes NOx", *Chemical Processing*, March 2004.

75. Nino, S. "Improve Material Balance in High-Purity Distillation Control", *Hydrocarbon Processing*, Process Control and Information Systems Bonus Report, November 2012.

76. O'Dwyer, A. *Handbook of PI and PID Controller Tuning Rules, Second Edition*, Imperial College Press, 2006.

77. Rinehart, N. and Ingram, N. "Smart Control Valve Diagnostics: Predictive Maintenance and Beyond", *Valve Magazine*, Vol. 16, No. 4, 2004.

78. Ruel, M. "Sticktion: The Hidden Menace", *Control*, November 2000.

79. Schuder, C. "Control Valve Rangeability and the Use of Valve Positioners", *Advances in Instrumentation*, ISA 26th Annual Conference, Chicago, Vol. 26, Part 4, Paper 817, October 1971.

80. Schork, E. et al. *Control of Polymerization Reactors*, Marcel Dekker, Inc., 1993.

81. Shinskey, G. *pH and pION Control in Process and Waste Streams*, McGraw-Hill, 1973.

82. Shinskey, G. *Energy Conservation Through Control*, McGraw-Hill, 1978.

83. Shinskey, G. *Controlling Multivariable Processes*, ISA, 1981.

84. Shinskey, G. "The Effect of Scan Period on Digital Control Loops", *InTech*, June 1993.

85. Shinskey, G. *Feedback Controllers for the Process Industries*, McGraw-Hill, 1994.

86. Shinskey, G. *Process Control Systems, Fourth Edition*, McGraw-Hill, 1996.

87. Shinskey, G. "PID-Deadtime Control of Distributed Processes" *Control Engineering Practice 9*, 2001.

88. Shinskey, G. "The Power of External-Reset Feedback" *Control*, May 2006.

89. Shinskey, G. "Multivariable Control of Distillation" *Control*, May 2009.

90. Shinskey, G. "Meditating on Disturbance Dynamics" *Control*, May2011.

91. Skogestad, S. "Simple Analytic Rules for Model Reduction and PID Controller Tuning", *Journal of Process Control*, Elsevier, 2003.

92. Skogestad, S. and Shamsuzzoha, M. "The setpoint overshoot method: A Simple and Fast Closed-Loop Approach for PID Tuning", *Journal of Process Control 20*, Elsevier, 2010.

93. Skogestad, S. and Downs, J. "An Industrial and Academic Perspective on Plantwide Control", *Annual Reviews in Control*, Vol. 35, Issue 1, Elsevier, April 2011.

94. Smith, C. "Watch Out with Variable Speed Pumping" *Chemical Processing*, May 2008.

95. Smith, C. *Practical Process Control – Tuning and Troubleshooting* Wiley, 2009.

96. Smuts, J. *Process Control for Practitioners*, OptiControls, 2011.

97. Spitzer, D. "Getting up to Smart Speed" Control, September 2008.

98. Spitzer, D. *Variable Speed Drives: Principles and Applications for Energy Cost Savings, Third Edition*, ISA, 2008.

99. St. Clair, D. *Controller Tuning & Control Loop Performance – A Primer*, Straight-Line Control Company, 1989.

100. Trevathan, V. *A Guide to the Automation Book of Knowledge* ISA, 2006.

101. Ziegler, J. and Nichols, N. "Optimum Settings for Automatic Controllers", *ASME Transactions*, 1942.

102. Ziegler, J. and Nichols, N. "Process Lags in Automatic Control Circuits", *ASME Transactions*, 1943.

Index

Printed and bound by CPI Group (UK) Ltd, Croydon, CR0 4YY
16/09/2026
14963426-0001